AF587894

HUMAN CHROMOSOMES AND AGING: FROM 80 TO 114 YEARS

Human Chromosomes and Aging: From 80 to 114 Years

Teimuraz Lezhava

Nova Biomedical
New York

Library of Congress Cataloging-in-Publication Data
Available upon request.

ISBN 1-60021-043-0

Published by Nova Science Publishers, Inc. ✢ New York

CONTENTS

ACKNOWLEDGEMENTS

I wish to express sincere thanks to Dr. Olivier Toussaint (Department of biology, University of Namur, Research Unit on Cellular Biology) for critical reading of the book and valuable recomendation

FOREWORD

The biological phenomenon of longevity is an area of great interest to gerontology specialists and the lay public. The first conference on aging convened by Bogomolets in 1940 was the starting-point, from which longevity research began to unfold. Learning the biology of the very senile human is essential for delineation of the species-specific longevity potential.

The book of T.A. Lezhava, a recognized geneticist and gerontologist, is the first major synthesis in the important and unexplored field of morphological and functional alterations of human chromosomes in very old humans. It is accepted by biologists that primary mechanisms of aging are related to impairment of genomic regulation, and this book is a major effort to provide proof of that. The book presents a comprehensive review of the morphology and function of chromosomes in long-livers. The wide range of topics includes cyclical chromosome properties, mutations, repair, progressive chromosome heterochromatinization with increasing age, roles of nucleolar organizer regions, sister chromatid exchanges, homolog relationships, heterochromatin regions and other chromosomal features in very old age. The author concludes from his analysis that heterochromatinization is a key determinant of genetic apparatus function during senescence and is an area to exanine when seeking life-prolonging interventions.

This book integrates progress in the field of the cytogenetics of aging and, of special importance, abundant and unique personal evidence. I am sure that this book will appeal to specialists and a wider readership of biologists and medical professionals with an interest in gerontology.

Academician V.V. Frolkis

PREFACE

This monograph deals with one of the chief problems of biology and medicine – the genetics of aging, in particular, the morphofunctional changes of human chromosomes during aging. It should be mentioned that the universality of the work resides in its general-purpose character, purposefulness and the material obtained from individuals aged from 80 to 114.

The analysis of the obtained results shows:

1. Chromosome progressive heterochromatinization (condensation of eu- and heterochromatin regions) occurs in aging;
2. Decrease of repair processes and increase of frequency of chromosome aberrations in aging are secondary to the chromosome progressive heterochromatinization. Chromosome heterochromatinization is a key factor of aging;
3. Chromosome heterochromatinization may be the reason for some senile pathology; Peptide bioregulators induce deheterochromatinization of chromosomes in old age.
4. Chromosome heterochromatinization is the area where one should seek solutions for the prolongation of the span of life.

This monograph is intended for gerontologists, geneticists and for a wide range of scientists working in the field of experimental biology and medicine.

INTRODUCTION

Aging is one of the most complex issues in biology. The aging process is a gradual deterioration of body functions. The deterioration becomes recognizable at the end of the reproductive span, as it merges into the aging period. The functional decline in humans is a linear process with insidious manifestations beginning at the age of 20 +/-10 (Mazin, 1994).

Gerontology is a science concerned with aging problems (*geron*, old man, *logos*, science). Biological gerontology deals with fundamental (molecular, genetic, biochemical, cytological, physiological) aspects of aging and its causal basis. Clinical gerontology (geriatrics) is the study and management of diseases of old age (cardio- and cerebrovascular ones, cancers, arthritis, rheumatism, autoimmune diseases). Social psychological gerontology is a branch related to psychology and social well-being of the elderly.

Experimental evidence gives rise to a multitude of concepts and models which seek to explain the aging process and basically falls into two categories.

The first category includes explanations of aging that are derived from changes at the genomic level. Concepts of the second category assert that aging occurs because of impairment of gene-determined secondary products (enzymes, collagen and hormones) and destruction of cellular structures, such as membranes, lysosomes, mitochondria. It is evident however, that this impairment is secondary since any change of the structure or regulation should be underlied by functional change of the genome which is responsible for the synthesis of various components (Kanungo, 1982). Therefore, it is appropriate to concentrate on genetic information levels since aging must occur in the genome (Hasty and Vijg, 2002).

Chapter I

Genetic Basis of Aging

General Considerations

Patterns of species-specific mortality provide the best argument for a genetic causation of aging. Pioneering studies of human aging by geneologist Bell (1918) demonstrated that life-spans of children of parents who died at ages below 60 years were almost 20 years shorter as compared to those of children whose parents lived longer than 80 years. Kallmann (1975) examined life-spans of 78 adult monozygotic and 102 dizygotic twins in New York state and found out that an average intrapair age difference showed considerable variability: from 36 months in the monozygotic twins to 74.6 months in dizygotic twins of the same sex; by contrast, the age difference of the dizygotic twins of opposite sexes was 106 months.

A longitudinal study in twins aged 60 years or more showed that monozygotic twins had a lesser difference of life-spans than dizygotic twins (Jarvik et al., 1960).

A similar correlation has been reported for survival of parents and children and twins grown up in different environments (Calow, 1978). This evidence also suggests a genetic basis of aging. Extensive statistical surveys of Lints (1978) and Asadov and Berdyshev (1986) have suggested a greater likelihood of longevity of individuals born by long-lived parents.

Representatives of a species require an equal time to attain sexual maturity. For example, the time is 10 weeks in rats and 12 years in humans. Likewise, the reproductive period has a comparable duration in species members. This suggests that developmental changes are genetically controlled and follow a genetic program. Another point of importance is that every species has a characteristic life-time. For example, the life-time is about 30 days for *Drosophila*, about 3 years for rats, and about 40 years for horses. A maximal human life-time is about 145 years (Mazin, 1994). In addition, there is a positive correlation between bees' lengths of life in a hive and honey outputs, another indication that a life-time is genetically determined (Milne, 1985). Maximal and average life-times within a species are largely dependent on environmental influences.

The maximal human life-time increased twofold during the last two-three million years. An average increment of it during 100,000 years was 14 years (Cutler, 1975). During the 100,000-year period, 40 to 250 genes might have sustained mutations that resulted in an adaptive amino acid substitution. This implies a modest number of mutations in structural genes and their scarce phenotypical effects on longevity. It is conceivable that a rapid evolution in the duration of primate life-times occurred due to the regulator gene change at the level of key enzymes and successive important contributions of the enzymes to morphological development and enlargement of the brain (King and Wilson, 1975) as related to chimpanzee and man. Mutations of the regulator genes controlling enzyme levels are crucial to animal development and may be a major determinant of human longevity (Hensler et al., 1994; Potapenko and Akifiev, 2000).

About 20 percent of the populations have genetic diseases that affect life expectancy, such as adult and juvenile diabetes, familial hyperlipidemia, sclerosis, alpha-1-antitrypsin deficiency, and cystic fibrosis. Genetic causes explain about 40 percent of infant mortality (Chieds, 1975). Multiple human genetic diseases amplify vulnerability of longevity genes to certain mutations (Takeshita et al., 1992; Martin and Oshima, 2000).

In multicellular organisms, aging begins as a decline in cell function and differentiation, and arrest of the cell division. This endowed evolutionary advantages and was fixed by selection together with indirect consequences of it: senility and death, and these in turn became useful adaptations for species. The genotype was given a developmental program (differentiation, growth, functioning), and triggering mechanisms of aging were built into this program (Tone et al., 1988; Luckinbill and Foley, 2000). That aging begins at conception and the life cycle is a continuous phase proved a misconception, as is indicated by evidence from experimental embryology and developmental biology. The common facts about programmed cell death during the embryogenesis make the best illustration that morphogenesis and aging are radically different biological processes (Goya, 1986; Hulbert, 2003). If aging is as natural as embryogenesis and development, effort should be undertaken to identify universal events that take place in cells of any organism regardless of its place in the evolutionary hierarchy and, probably, regardless of the cell's environment. On the other hand, it is thought that aging is a function of intracellular processes since the cell is a source of all regulatory mechanisms and energy production (Hahn, 1971; Hornsby, 2002).

Variability of organ, tissue and cell aging rates suggests that functional failure of certain cell types is a fundamental cause of aging (Hayflick, 1985, 2000). It is known that cellular regulation and utilization and transmission of genetic information are maintained by the central information matrix of the genome and nuclear control systems. Given that cell aging is a universal phenomenon, it is the central control level where we should look for a major universal aging-related 'damage' which induces secondary destructive changes in the cytoplasm (Hahn, 1971). Serial studies of Prescott et al. (1972) and Wright and Hayflick (1975) examined the nucleus and cytoplasm for a 'clock' which might commit a cell pool to doubling. Treatment of cells with cytochalasin B and centrifugation yielded enucleated cells (cytoplasts) that remained viable for many days. Young cytoplasts were hybridized to old cells with inactivated Sendai virus, or a reverse procedure was used to obtain heteroplasms. Counts of cell population duplications induced by the young and old cytoplasm revealed that the 'clock' resided in the nucleus.

Culture aging has been evaluated in the studies of Reff and Schneider (1981), Muggleton-Harris and Defaria (1985), Dvalishvili (1989), Lezhava and Dvalishvili (1992) employing biochemical, clonal and genetic assays and analysis of passages, clones and sister chromatid exchanges in human skin fibroblast and blood cultures. Cell culture aging in vitro showed a relationship to in vivo cell aging.

According to Van-Ganzen (1979), there are three arguments for a complete analogy between in vitro and in vivo aging of cells: (1) proliferative ability of cell strains from every animal species is inversely proportional to the age of cell donors; (2) there is a good correlation between the duration of a species life span (human, mouse, mink, hen, Galapagos tortoise) and the number of fibroblast mitoses in the culture; and (3) human fibroblasts poorly proliferate in the culture if they are isolated from patients with progeria, a syndrome associated with very premature aging. In vitro cell aging is recognized through impairment and eventual loss of the proliferative ability of the culture.

A fundamental issue in gerontology is whether ideal conditions for functioning and replication can protect eukaryotic cells from aging and death inevitable for an animal donor of the cells. Another major problem is the elucidation of whether in vitro studies use normal or abnormal cells. It has been established that aging occurs in a normal cell population. Since behavior of normal cells in vivo correlates with those in vitro populations, the latter should be normal too. Studies concerning this problem must not use 'immortal' lines descending from HeLa and L cells because they have one or two abnormal traits (Hayflick, 1979).

Studies of cultivated normal diploid fibroblasts from human embryos (Hayflick and Moorhead, 1961; Hayflick, 1965) demonstrated that, unlike transformed heteroploid cell lines, initial populations of the normal fibroblasts experienced a limited number of serial duplications and then degenerated and died. Normal diploid fibroblast cultures displayed three developmental phases: a primary culture of freshly isolated cells from an animal donor; a growing culture of proliferating cells and a culture of senescent cells with nearly arrested divisions. It was found out that with the best of cultural conditions, death of the human embryonic fibroblasts was inevitable after 50 (±10) duplications of the initial population (the third-phase phenomenon), and the numbers of generations varied with employed strains of primary cells. Death of the cultivated cells after a number of duplications proved to be their hereditary characteristic. Experimental support for this evidence has been provided in numerous laboratories (Martin et al., 1970; Goldstein, 1974; Schneider et al., 1977; Goldstein, 1990; Cristofalo and Pignolo, 1993; Howard, 1996) using cells from different species for culture and variable environmental and cultural conditions (Hayflick, 1979; Warner et al., 1992; Campis, 2001).

Further studies have shown a correlation between cell survival in the culture and donors' ages (Hayflick, 1979). The proliferative ability of adult fibroblasts was found to be lower (14 to 29 population duplications in 8 adult donors), as compared to human embryonic fibroblasts (35 to 63 duplications in 13 embryos) (Hayflick, 1979, 1984). The q arm of human chromosome 1 carries a gene or set of genes which are altered in the cell lines assigned to complementation group C and is involved in the control of cellular senescene (Hensler et al., 1994).

Likewise, survival of graft cells has been correlated with ages of tissue grafts (Harrison, 1975). Moreover, Hayflick (1975, 2000) found that human fibroblasts retained their division potential after 16-year storage in liquid nitrogen at –196 °C. The investigator thawed 130 ampoules (one ampoule was being thawed for a month) (WI-38), and the number of fibroblast divisions in each one was invariably equal to the prestorage count (50 ± 10 duplications of an initial population).

In summary, available studies indicate that aging is a genetically controlled physiological process and in vitro cell aging correlates with in vivo aging.

THEORY OF AGING

Of the multiple theories of aging (Hayflick, 1985; Medvedev, 1990; Morris, 1994; Harman, 1999; Luckinbill and Foley, 2000; Gladyshev, 2001; Lezhava, 2001a; Vijg and Dolle, 2002; Warner and Sierra, 2003; Kappeler and Epelbaum, 2005), the most widely accepted one is the molecular-genetic theory. Three offshoots of this theory are **programmed aging**, **stochastic aging**, and **repair impairment**, although they differ in approach rather than essence.

The first concept states that any organism has a genomic **program of aging** which is fulfilled at the last phase of its life. Annual plants, specialized embryonic cells and numerous insects have been cited by Cutler (1978) as examples of organisms with genetically programmed death. The life cycle of these organisms appears to be terminated by one of the following mechanisms: (1) turning on of specific killer genes (most likely); and (2) turning on or limiting of essential metabolic products and/or life-support processes. Aging is the result of genome-programmed degeneration (Mier and Kerkhof, 1990; Claire et al., 1994; Lezhava, 2001a; Hulbert, 2003; Huber et al., 2003; Yang et al., 2005).

Rockstein et al. (1977) have pointed out that maximal life-times of most species are genetically determined. Thus, the life-time is about one day for May flies, 100 days for fruit flies, 1 year for shrews, 10 years for rabbits, and 100 years for humans. The implication arises that 'programmed genes' for aging must exist.

The main idea of the **stochastic aging theory** is that cells are exposed to environmental factors, and some of the adverse sequelae cannot be reversed. This results in accumulation of a lethal amount of cellular lesions. Stochastic aging is thought to be an outcome of incidental events (somatic mutations, transcriptional, translational or cell-cycle errors) (Kirkwood, 1988). Orgel (1963, 1970) proposed the **error catastrophic concept** which defined aging as accumulation of 'erroneous macromolecules' (DNA, RNA, proteins) in the cell's genetic system. Errors in protein synthesis build up until catastrophe, i.e. cell death, occurs. Orgel attempted to circumvent the issue of lesion accumulation by arguing that a positive feedback is available to promote the repair of deformed molecules. It has been suggested by Calow (1978) that even with the repair this feedback may work only with the proviso that errors remain localized to a certain level. Orgel supplemented his concept with a definition of proportional relationship (α) between the synthesis errors and newly produced proteins.The errors lead to catastrophe only when α is above unity; when it is not, the amount of the errors remains stable and no catastrophe will ensue. If the proportional relationship becomes increasingly linear, a protein-producing system grows more efficient and resistant to failure; therefore, the error mechanism alone is unlikely to account for aging.

Remacle (1985) has documented in vitro accumulation of changed enzymes because of an aging-related decrease in the cellular metabolism.

Stochastic explanations of aging that define lesions as mutations are referred to as the **somatic mutation theory.** An overview of experimental findings (Martin, 1979) suggests that (1) mutations do accumulate in various somatic tissues of senescent mammals; (2) the genetics of the mutation accumulation is consistent with life-spans of mammalian species (Martin, 1979; Lee, 1999).

Most researchers believe that progressive accumulation of somatic mutations may be a major factor in the age-related degeneration. According to the mutational aging theory of Burnet (1974), genetic loci (40 to 50) are responsible for the rates of somatic mutations and their sequelae. For a mutation to occur, a primary mutant cell must give rise to a clone of mutant descendants. This may be produced in three ways. First, the onset of a somatic mutation in an embryonic cell. Second, a primary somatic mutation entails a cascade of secondary mutations with subsequent elimination of involved clones, the situation the Orgel error catastrophe concept predicts to happen when a gene mutation prevents key enzymes from efficient maintenance of accurate RNA replication and repair. Third, neoplastic growth is associated with two or more somatic mutations. Genetic instability as a determinant of aging may incur an increasingly greater accumulation of errors in messenger molecules. Progressive error accumulation in postmitotic or actively dividing cells might act as a clock. This would induce secondary errors resulting in aging manifestations (Hayflick, 1979, 1985; Vorobtsova et al., 2001). The frequency of somatic-cell chromosome breakage was significantly increased in neuroblasts of D. melanogaster males with somatically active P elements in the presence of P [ry+delta 2-3] (99B) transposase (Woodruff and Nikitin, 1995).

Several recent studies have determined a remarkable function for the human SIRTI protein, which is the closest human homolog of yeast Sir2. SIRTI specifically associates with the p53 tumor suppressor protein and deacetylates it, resulting in negative regulation of p53-mediated transcriptional activation. Importantly, p53 deacetylation by SIRTI also prevents cellular senescence and apoptosis induced by DNA damage and stress (Smith, 2002).

One of stochastic versions of aging focuses on **free radical** damage. Free radicals are chemical species that contain an impaired electron in an outer orbit. This electron makes them highly reactive. Free radicals are produced as transient intermediates in the course of normal metabolism, e.g. during mitochondrial oxidation. The reaction of a free radical with a stable molecule produces another radical, and this often results in a chain reaction, in which the single free radical initiates a process that consumes many stable molecules. It has been suggested that some free radicals of a metabolic or spontaneous origin may contribute to the aging process. Since free radicals may attack important molecules, such as DNA, proteins and lipids, and since they also tend to be self-propagating, they are capable of generating considerable damage (Lamb, 1977; Gille, 1990).

The **free radical theory** postulates that generation of reactive oxygen species (ROS) radicals from HO, O2- and H2O2 by the Fenton reaction inducing DNA damage potentiate adverse exposures of life or aggravate diseases, thereby contracting the life-time. The concept has three major theses: (1) free radicals are exceptionally reactive; (2) oxidized free radicals, especially superoxide oxygen anions, are permanently produced in the body, providing a protection against microorganisms, but also posing some danger at molecular and cellular levels (Wickens, 2001; Akman et al., 2002; Kawanishi et al., 2002). The free radical concept explains that aging is associated with an imbalance between the production and control of free radicals, and these result in excessive free radical production and functional disorders at molecular, cellular, tissue and organ levels (Crastes de Paulet, 1990; Beckman and Ames, 1998). Denham Harman, the chief advocate of the free radical theory (Harman, 1956), indicates that principal sources of free radicals are the mitochondrial respiratory chain, phagocytosis, prostaglandin synthesis, and cytochrome P-450. Environmental and genetic factors may trigger 'free radical' diseases, such as cancer, atherosclerosis, Alzheimer's and Parkinson's diseases, hypertension, cataracts, diabetes, amyloidosis, immune deficiency of aging, etc. It has been suggested that life expectancy may be improved by 5 to 10 years by appropriate diets supplemented with antioxidants (α-tocopherol, carotinoids, heme-containing peroxidases, selenium-dependent glutathione peroxidase, superoxide dismutase, agents that elevate serum uric acid, ionol, 2-mercaptoethylenamine) (Harman, 1988). A general model is put foreword to explain the mechanism by which age-associated aneuploidies are produced. This is based on the free radical theory of aging, which assumes a rise in oxidative stress with age (Tarin, 1995; Harman, 1999; Gusev, 2000; Barja, 2002).

It has been experimentally demonstrated that ionizing radiation, free radicals and aldehydes may interact with proteins and DNA to produce cross-linkages within or between their molecules. Malonic aldehyde and formaldehyde, related with the superoxide radical system, have a special potential to accumulate in cells in old age and to cause damaging effects.

The Theory of Stress. Many proliferative cell types like lung and skin human diploid fibroblasts, human melanocytes, endothelial cells, human retinal pigment epithelial cells, exposed to subcytotoxic stress (UV, tert-butylhydroperoxide, H_2O_2, ethanol, mitomycin C, hyperoxia, γ-irradiations, homocysteine, hydroxyurea) undergo stress-induced premature senescence (SIPS) in vitro (Toussaint et al., 2002).The appearance of SIPS could be due to exacerbated modifications of a limited number of parameters that also undergo, to a more limited extent, age-related changes, among multiple other age-related changes. Irreversible changes in gene expression take place when SIPS becomes established. Some genes become permanently underexpressed while other become overexpressed.A change in the cellular targets under positive feedback is operated by the establishment of cascades of new regulatory loops, eventually locking the system in a new attractor. Common and different pathways are induced after exposure to different kinds of subcytotoxic stress, changing the level of expression of common and different genes (Toussaint et al., 2001).

Adherents of the cross-linkage theory hypothesize that aging results from accumulation of macromolecular linkages and secondary disturbance of the intracellular transport (Bjorkstein and Tenhu, 1990). Bjorkstein (1977) suggested that the incidence of free radical-induced cross-linkages in vital macromolecules (protein and DNA) must increase with advancing age, interfering with their normal function. The cross-linkage may produce somatic mutation and impairment of DNA replication. It is known that DNA binding to histones is stronger in old cells.It was thought to be related to extended intermolecular linkages, a danger to transcription (Hahn, 1970).

Over time, postmitotic cells accumulate a non-degradable intralysosomal substance, lipofuscin, which forms due to iron catalyzed oxidation /polymerization of protein and lipid residues. **Lipofuscine** is often considered a hallmark of aging showing an accumulation rate that inversely correlates with longevity (Terman and Brunk, 2004). The potential contributions of free radical-mediated modifications to protein structure/function and alterations in the activities of two major proteolytic systems within cells, lysosomes and the proteasome are due to the age-dependent accumulation of lipofuscin.

There is immunochemical and ultrastructural evidence demonstrating the occurrence of a fluorescent protein cross-link derived from free radical-mediated reactions within lypofuscin granules of rat cerebral cortex neurons (Szweda et al., 2003).

Evidence that the 'classic' concepts of aging as accumulation of stochastic dna lesions require a serious update. A more comprehensive mutation theory is presently evolving to take into account genetic and epigenetic contributions to aging (Kirkwood, 1988).

An interesting hypothesis in favor of **regulation mechanisms of aging** has been proposed by Frolkis (1986). This investigator calls it the adaptation-regulation theory. This concept defines aging as an intrinsically controversial process, in which metabolic and functional decline (destruction) coexists with acquisition of important adaptations. With this combination of events, the homeostasis may long remain intact despite an impaired reliability or disorders of self-regulation mechanisms. The concept also contends that aging occurs because of primary abnormalities in regulator genes, successive impairment of metabolic pathways of selected proteins and, as a result, serious functional disorders of cells that are involved in protein synthesis. The onset of these abnormalities in neural centers, especially the hypothalamus, leads to major changes in the neurohumoral control and eventual failure of tissue metabolism and function.

Repair. Steric configurations of key enzymes mediating DNA repair is a crucial factor in the genetic control of aging (Friquet, 2002). There is evidence to indicate that the enzyme system responsible for the repair of DNA lesions is less efficient in senescent fibroblasts and fibroblasts derived from premature senescene syndromes (Little, 1976; Mikhelson, 1986; Vijg and Knook, 1987; Melaragno and Smith, 1990; Weirich-Schwaiger, 1994; Shen et al., 2003). This author (Little, 1976; Shen et al., 2003) has suggested that the repair and substitution process perpetually occurs during aging. Therefore, the accumulation of damage with increasing age may be induced either by accelerated rates of damage or a lower accuracy of repair. Since the latter is more likely, the accumulation of random molecular lesions is predictable during aging. Calow (1978) interpret aging as amplification of damage owing to compromized repair mechanisms. Intano et al. (2003) gave conclusive evidence that the mice germ line short-patch base excision repair activity appears to be relatively constant throughout spermatogenesis in young animals, limited by uracil-DNA glycosylase and DNA ligase in young animals, and limited by apurinic/apyrimidinic endonuclease in old animals.

Therefore, repair appears to be impaired in the course of aging. Although old cells are still capable of repairing DNA damage, repair rates are inferior to those of young cells.

Of the numerous theories of ontogenesis, the most plausible explanations of aging are offered by the current genetic hypothesis which describes aging as summary effects of what is defined as programmed aging, mutations, and impairment of repair mechanisms.

Cutler (1978) has concluded from his long-term studies that aging of mammals is evolutionarily underlied by actions of pleiotropic genes, both beneficial and deleterious. Since a greater part of the genome appears to be regulatory rather than structural genes,

Cutler presumed that the accumulation of structural chromatin lesions may affect the regulatory genes, resulting in abnormality of gene regulation, not synthesis of anomalous proteins. It may in turn cause an excessive loss of the genetic control at the cellular level and regulatory strategies at the systemic level. Such a dysfunction of chromatin in vital tissues might prove to be the main cause of phenotypical aging of mammals. Using molecular and biochemical approaches, important changes in gene expression and function of cell-cycle-associated products have been identified (Goletz et al., 1994).

Martin (1991; Martin and Oshima, 2000) has stated that the aging phenotype may be controlled by relatively few genes (7 to 70). Mutations of these genes may induce premature phenotypical aging which has the collective name of segmental progeroid syndrome and comprises ten syndromes: Down's, Werner's, Cockayne's syndromes, progeria, ataxia-telangiectasia, Sapes' syndrome, cervical lipodysplasia syndrome, Klinefelter's, Turner's syndrome, and muscular dystrophy syndrome. Three of these syndromes are chromosomal disorders presenting as quantitative and regulatory abnormalities rather than qualitative enzyme impairment. According to Martin (1979), control of aging is driven mostly by regulatory genes. Abnormalities of these genes are believed to be a sequela of chromosomal disorders that mainly define life-spans of mammals and occur in premature senility syndromes, such as Down's syndrome, a condition associated with 47, XX or XY, 21+ trisomies produced by a small extra autosome. Patients with Down's syndrome may display translocations and chromosomal disorders of a mosaicism type, when cells with the normal set of 46 chromosomes coexist with cells having trisomy 21 (Barenfeld, 2002). Patients with Turner's syndrome lack one X chromosome (Tumilovich and Lezhava, 1968; Cortes-Gutierrez et al., 2003), while those with Klinefelter's syndrome have an extra X chromosome (Warwick et al., 2003). These genetic entities do not affect any genes structurally, but they do produce disproportions of normal genes, obviously with disturbance to the gene regulation.

Chromatin

Higher organisms carry the genetic material in chromatin (chromosome), which is a complex nucleoprotein structure. Chromatin dynamics and higher order folding may be the key regulators of not only transcription but of all DNA-dependent processes in the nucleus (Nemeth and Langst, 2004).

Chromatin is organized into a granule of v-nucleosomes that are made of DNA with protein complexes (Olins and Olins, 1974). Chromatin has histone and nonhistone proteins. Histones make up octamers, or two molecules of each of the four histone proteins: H2A, H2B, H3, and H4. DNA is wound over them (about 160-200 nucleotide pairs per globule). The histone proteins interact via hydrophobic fragments. DNA complexes with the histones are separated by spacers that consist of about 40 base pairs, with the H1 class of 'linker' histone bound to them. (Nakagama and Takami, 2001; Chikhirzhina and Vorob'ev, 2002).The nucleosome filament is formed by connecting nucleosome cores with linker DNA of variable length and by association of one histone H1 per nucleosome. The nucleosome filament is condensed further into compact 30 nm fibres and higher order structures. The structures are present in interphase and metaphase and are the target for damage formation and repair. Transcriptional activation has been associated with changes in the structure of both chromatin and nucleosomes (Herrera et al., 2000). These changes are mediated by chromatin remodeling complexes and by reversible modification of histone (Wolffe and Hayes, 1999; Vaquero et al., 2003).

The functional state of chromatin can be regulated by at least three different strategies: differential association of non-histone proteins, covalent modification of the histones themselves and ATP-dependent mobilization of nucleosomes. The latter, energy-dependent alterations are brought about by so-called chromatin-remodeling factors, multiprotein complexes containing ATPases of the SWI2/SNF2 subfamily of the SWI/ SNF protein complex. Biochemical analyses of purified yeast SWI/SNF complex and its human counterpart demonstrated that the machinery can modulate histone-DNA interactions such that the accessibility of nucleosomal DNA is much enhanced, facilitating the interaction of proteins with their binding sites on DNA.The SWI/SNF complex and related human BRM and BRG1 complexes have since been shown to be involved in many important processes involving chromatin substrates, such as transcription that leads to cell cycle progression, cellular differentiation, replication, recombination and repair (de La Serna et al., 2001; Langst and Becker, 2001; Ura et al., 2001).

Studies of chromatin compaction in response to changes in the ionic environment have established that the phenomenon can be accounted for by electrostatic interactions between DNA, histone proteins, and free ions. Major contributions to these interactions are provided by the N-terminal domains of the core histones, which contain ~50% of the basic amino acids of the octamer, and the C-terminal domains of the histones, which contain ~ 60% of the positive charges in these molecules. Indeed, chromatin compaction requires the presence of the core histone N-termini and 30-nm chromatin fibers are not formed in the absence of linker histones (Bednar et al., 1998).

Nonhistone proteins include metabolically involved enzymes; these proteins are responsible for chromatin structures and act as activators or repressors. Half of any nonhistone protein consists of actin, myosin, tropomyosin, and tubulin. The nonhistone proteins are found both in active and inactive portions of the genome. They are highly variable in higher organisms, making 10 to 60 percent of total proteins. The nonhistone proteins are different at various phases of cell differentiation and in different organs. They are directly responsible for regulation of the gene activity (Chikhirzhina and Vorob'ev, 2002).

To serve their function of coding and use of genetic information, chromosomes are structurally and functionally differentiated into condensed and decondensed regions. Therefore, they must have a transcription regulator mechanism based on chromosomal modification (condensation and decondensation of individual genes or whole chromosomes) during cell differentiation, and specific functions.

Euchromatic or decondensed active regions contain the bulk of nuclear genes, i.e. chromosomal strand sites that differentially control traits of an organism. The chromomer in strand of these regions experiences unspiraling in the quiescent nucleus. The euchromatin regions have a high biochemical differentiation, and the loss of a tiny particle by these regions is disastrous for the cell. Unlike the euchromatin regions, heterochromatic, or condensed regions remain spiral during most of their cycle and produce massive bodies even at the interphase. The heterochromatic regions have a poor biochemical differentiation, and the cell survives even after a considerable loss of the heterochromatic regions. Cytogenetic studies have shown that the heterochromatic regions exert strong impacts on adjacent euchromatic regions by modifying their cycles and inducing their persistent heterochromatinization. As a result, dominant traits controlled by loci of a given euchromatic site are usually converted to recesive ones; this means that condensation-heterochromatinization of an euchromatine site inactivates the genes it contains (Prokofyeva-Belgovskaya, 1986; Cremer et al., 2000; Carvalho et al., 2001).

A differentiated cell achieves a stable maturity with unfolding life cycle of an organism. Only a small portion of its inherited genome is actively involved in protein and RNA synthesis. Most of normal homeostatic change, especially age-related loss of cell function, occurs in the chromosomal domains. This consideration should be of major account in concepts that explain cell aging by genome damage. Possible development of this process in euchromatin makes the event more probable.

Research from recent years suggests that, despite the abundance of studies on the genetics of aging, numerous issues related to age-specific changes of nucleoprotein structures (chromosomes) still remain controversial or obscure. Important cues to chromosome function at late developmental phases and diseases of senility may be obtained by cytogenetic evaluation of chromosome structure and function: cyclical properties, condensed eu- and heterochromatic regions, denaturation parameters of total heterochromatin, relations of homologs to chromosome, mutations (aneuploidy, aberrations), nucleolar organizer regions, unscheduled DNA synthesis and patterns of sister chromatid exchanges in very old individuals.

Chapter II

Mutation

Aneuploidy

Aneuploidy in Peripheral Lymphocyte Cultures

One of initial attempts to correlate chromosome changes in lymphocyte cultures with age was undertaken by Jacobs et al. (1961). Spontaneous aneuploidy was examined in 97 apparently healthy subjects (50 men, 47 women) of 5 to 82 years of age. The findings first indicated that variability of the diploid chromosome number in somatic cells from the normal individuals is related to age: the incidence of aneuploid cells increased with age from 5.21 percent (at ages from 25 to 34 years) to 13.7 percent (at ages older than 65 years). The higher incidence was determined by the increase in hypodiploid cells (4.02 to 8.44 percent) and hyperdiploid cells (0.8 to 3.1 percent). The authors suggested that age might have a contributary role in addition to artifacts.

That the incidence of aneuploidy increases with age was confirmed by Jacobs et al. (1963) in a greater population (8380 cells from 247 subjects in the age range from the neonatal period to 75 years or more). The incidence was 3.66 percent at ages below 14, 5.64 percent at 15 to 64 years, and 12.06 percent at ages above 65 years. Females showed a loss of 6-X-12 group chromosomes and males lost 21-22-Y group chromosomes.

Similar results were obtained by Hamerton et al. (1965) in a study of 4910 metaphases from 200 subjects of both sexes aged 1 to 82 years (two populations). The incidence of aneuploidy rose with age in the case of loss or excess of X-chromosome in female cells and Y-chromosome in male cells. The onset of aneuploidy was interpreted as a result of mitotic disturbance that are usually associated with loss or excess of any chromosomes and eventual death of cells. However, the cells without genetically inactivated X or Y chromosomes survived.

Increased pools of aneuploid cells were found in lymphocyte cultures of elderly women with loss of C group chromosomes. The pools were greater than those of 19- to 21-year-old girls and newborn infants (Goodman et al., 1969). Hypodiploidy accounts for the increase of aneuploid cell counts in aging men and women (Cadotte and Fraser, 1970).

Kerkis et al. (1967) evaluated the incidence of aneuploidy in 3482 metaphases of 73 subjects of different ages and found out that (1) percentage of aneuploid cells in peripheral lymphocyte cultures increased with age from 0.30 percent (at ages under 5) to 10.27 percent (at ages older than 60); (2) C group aneuploidy appeared to be induced by the X chromosome in women and by both the X and Y chromosomes in men; (3) hypodiploid cells were a primary determinant of the age-related increase in aneuploidy incidence.

The incidence of cells with aneuploid chromosome sets was found to vary with age and sex. Court Brown (1966) examined chromosomes of 12,308 cells from 438 subjects (207 men, 231 women) aged 15 to 90 years. A mean incidence of the aneuploid cells was 7.31 percent. The proportion of the aneuploid cells was 6.36 percent in 15- to 64-year-old women (hypodiploid, 5.75 percent; hyperdiploid, 0.61 percent) and 12.32 percent in women aged 65 years or older (hypodiploid, 11.01 percent; hyperdiploid, 1.31 percent). The incidence of the aneuploid cells in men aged 64 years or older was 4.24 percent (hypodiploid, 3.88 percent; hyperdiploid, 0.36 percent); men aged 65 years or older showed 6.35 percent of the aneuploid cells (hypodiploid, 5.76 percent; hyperdiploid, 0.59 percent). The incidence of aneuploidy in women older than 65 was well above that in men of respective ages.

Sandberg et al. (1967) assayed chromosomes of 10,393 cultured lymphocytes from 171 subjects (99 women, 72 men) aged 75 years or older. The age was found to increase the incidence of hypodiploid cells in women, but not hyperdiploid ones.

Jarvik and Kato (1970) evaluated chromosomes of 61 twin pairs (23 men, 38 women) of 77 to 93 years of age; 30 individuals aged 15 to 17 years were a control group. Again, elderly women exhibited a high incidence of hypodiploid cells, while men of comparable ages did not. Loss of C group chromosomes was more common in elderly men, as compared to that in young men.

However, a number of studies did not reveal a relationship between aneuploidy and age. Aneuploidy of chromosomes were examined by Kleisner de Galan (1966) in 10 apparently healthy subjects (5 men, 5 women) aged 63 to 91 years. Assaying of 564 cells showed that chromosome counts were comparable to those seen in a middle age. Bloom et al. (1967) reported their findings in lymphocyte cultures from 30 women aged 70 years or older who had experienced the atomic bombardment of Hiroshima and Nagasaki. The incidence of aneuploidy (and specifically hypodiploidy) in these old women did not significantly differ from that in young women. Kadotani et al. (1971) examined blood culture metaphases in 49 couples of 30 to 79 years of age who were permanent residents of Hiroshima and had at least three children. No statistically significant age- or sex-related variability of aneuploidy was found in 4,425 male and 4,516 female cells. The incidence of cells with a modal chromosome count in men and women was 93.62 and 94.18 percent, respectively, and the incidence of tetraploid cells was 1.24 and 0.71 percent.

The study of El Zawahry and Khishin (1980) has evaluated lymphocytes from 78 donors of 5 to 68 years of age using differential quinacrine staining. Subjects of both sexes did not display a significant change of hypodiploidy incidence with advancing age.

Nonetheless, numerous studies did identify a statistically significant increase of aneuploid cell percentages (commonly hypodiploidy) which was directly proportional to age. Interphase and metaphase assessment in lymphocyte cultures from very old persons demonstrated that increased percentages of aneuploid cells were commonly related to deficient C group chromosomes in women (Kuznetsova, 1972; Jarvik et al., 1974, Abruzzo et al., 1985; Richard et al., 1994; Hando et al., 1994; Catalan et al., 1995) and G group chromosomes in men (Jarvik et al., 1974; Guttenbach et al., 1994; Nath et al., 1995; Robbins et al., 1995). The study of Fitzgerald (1975) employed the method of G banding in the evaluation of lymphocyte cultures of women from 60 to 87 years of age. The study gave accurate evidence that increased metaphase percentages of cells with hypo- and hyperdiploid chromosome sets relative to controls were due to a deficient or excessive X chromosome. Fitzgerald first observed that premature centromere separation in one or both X chromosomes in women older than 59 occurs four times as frequently (in 2 percent of metaphases) as in women younger than 40. The early centromere separation was interpreted as a cause of chromosome nondisjunction in lymphocyte cultures of senile women. The centromere separation in senile women was found to be earlier in chromosomes X, 8 and 10 and later in chromosomes 2, 3, 5 and 18 (Bajnoczky and Mehes, 1985, Mukherjee and Weinstein, 1988). Significant age-dependent difference was also noted for autosomes 27.5% among the older women (above 50 years old) and 20.3% among the younger women (below 30 years old) (Catalan et al., 1995).

A positive correlation between donors' ages and the incidence of hypodiploidy has been identified as Martin et al. (1980) evaluated aneuploidy lymphocyte cultures of patients with senile dementia aged 65 years or older (111 women, 34 men) and normal subjects (31 women, 32 men) aged 18 to 32 years. In women, hypodiploidy was commonly underlied by loss of X chromosomes and C group chromosomes. Hando et al., (1994) were obtained blood samples from 8 newborn females and 38 adult females ranging in age from 19 to 77. Isolated lymphocytes were cultured according to standard techniques and blocked with cytochalasin B. Two thousand binucleated cells per donor were scored using a modified micronucleus assay to determine the kinetochore status of each micronucleus. Slides were then hybridized with a 2 kb centromeric X chromosome-specific probe labeled with biotinylated dUTP, and detected with fluorescein-conjucated avidin. All micronucleated cells were located and their X chromosome status was determined. These results show a significant increase with age in the number of micronuclei containing an X chromosome.

Mattevi and Salzano (1975a) undertook a study of chromosome changes in lymphocyte cultures of 60 subjects from 62 to 96 years of age and 60 control subjects from 10 to 13 years of age. Of 3,900 cells, 3,600 were from three-day and 300 from two-day cultures. The findings indicated that the incidence of aneuploidy and structural chromosome damages increased with advancing age. Furthermore, elderly women showed an increased incidence of cells lacking a group C chromosome, while no chromosome loss was seen in elderly men.

The aneuploidy study of Jarvik et al. (1976) followed up old subjects at 6-year intervals and showed that:

(1) The aging process is associated with chromosome loss; the incidence of cells with hypo- and hyperdiploid chromosomes increases with age;
(2) C group chromosome loss in old women is far more common as compared to other chromosomes;
(3) Aging women have higher percentages of cells with hypodiploid chromosome sets and tend to lack individual chromosomes in the metaphase periphery. Geigl

et al. (2004) show an association between age - related aneuploidy and the gene expression level of genes involved in centromere and kinetochore function.

A total of 1,000 lymphocyte interphase nuclei per proband from 90 females and 138 males aged from 1 week to 93 years were analyzed by in situ hybridization for loss of the X and Y chromosomes, respectively (Guttenbach et al., 1995). Both sex chromosomes showed an age-dependent loss. In males, Y hypodiploidy was very low up to age 15 years (0.05%) but continuously increased to a frequency of 1.34% in men aged 76-80 years. In females, the baseline level for X chromosome loss is much higher than that seen for the Y chromosome in males. Even prepubertal females show a rate of X chromosome loss, on the order of 1.5%-2.5%, rising to approximately 4.5%-5% in women older than 75 years. Dividing the female probands into three biological age groups on the basis of sex hormone function (<13 years, 13-51 years, and >51 years), a significant correlation of X chromosome loss versus age could clearly be demonstrated in women beyond age 51 years. Females age 51-91 years showed monosomy X at a rate from 3.2% to 5.1%. In contrast to sex chromosomal loss, the frequency of autosomal monosomies does not change during the course of aging: chromosome 1 and chromosome 17 monosomic cells were found with a constant incidence of 1.2% and 1% respectively. These data also indicate that autosome loss in interphase nuclei is not a function of chromosome size.

Neurath et al. (1970) evaluated 7,067 metaphases from cells of 139 apparently normal subjects aged two weeks to 93 years for a relationship between age and chromosome number. The conclusions were (1) chromosome loss is a function of chromosome size in all chromosome groups (A to G); (2) an unknown factor must exist to account for the loss of G group chromosomes or Y chromosomes; (3) loss of some chromosomes may be age and sex-related. Nath et al., (1995) investigated the status of kinetochore proteins, the relationship between Y chromosome loss and increased micronucleus formation with age. Umbilical cord blood samples were obtained from 18 newborn males, and peripheral blood was obtained from 35 adult males ranging in age from 22 to 79 years. Isolated lymphocytes from all 53 donors were cultured and blocked with cytochalasin B. Two thousand binucleate cells per donor were scored using a modified micronucleus assay to determine the kinetochore status of each micronucleus. This assay showed 23.8% of the micronuclei to be kinetochore-positive, while 76.2% of the micronuclei were kinetochore-negative. Cells were then hybridized with a 3.56-kb biotinylated Y chromosome-specific probe. All micronucleate cells were relocated and their Y probe status was determined. A significant increase in Y-bearing micronuclei with age was observed. Metaphase cells from the same samples were analyzed for the presence or absence of the Y chromosome. The relationship between Y chromosome-positive micronuclei and Y chromosome-negative metaphase cells was highly significant, suggesting that Y chromosome-deficient metaphase cells result from cells which had previously lost a Y chromosome due to micronucleation. The cause of micronucleus formation from a lagging Y chromosome appears probably to be either a faulty or a diminished amount of kinetochore protein. The age-associated increase of sex chromosome loss was studied using fluorescence in situ hybridization (FISH) on interphase nuclei (Bukvic et al., 2001). The loss of Y signals was observed in approximately 10% of interphase cells from the centenarian males, that is six times more often than in the younger control men (approximately 1.6%).The frequency of X signal loss (approximately 1.7%) in young women was similar to that observed in male controls of the same age but the incidence of the X chromosome aneuploidy in centenarian females was appreciably higher (approximately 22%) than that found for the Y chromosome in males.

Studies of aneuploidy in cells of the bone marrow (O'Riordan et al., 1970; Pierre and Hoagland, 1972; Harrison, 1975; Shekhter-Levin et al., 1992), ovaries (Fang et al., 1975) and sperm (Martin and Rademacker, 1987; Martin, 1989) of old individuals have demonstrated a tendency of increased Y chromosome loss by marrow cells and X chromosome loss by marrow and ovarian cells. But in sperm the incidence of structural chromosomal aberrations strongly correlated with age, while no correlation was seen between aneuploidy and age. The method of simultaneous three-chromosome FISH was developed using repetitive DNA sequence probes for chromosomes 8, X and Y and applied to semen of 14 men from two healthy groups who differed in their average ages (46.8 +/- 3.1 years, n = 4; 28.9 +/- 5.0 years, n = 10) (Robbins et al., 1995). The frequencies of disomic sperm determined by FISH compared well with frequencies obtained using the hamster-egg technique for human-sperm cytogenetics and with the frequencies of disomic and diploid sperm reported in previous FISH studies in this laboratory. The older group had statistically higher frequencies of sperm carrying sex chromosomal disomy than the younger group (Robbins et al., 1995).

Overall, evidence from cytogenetic studies (Table 2.1.1-1) suggests that most of old individuals had a significant increase in percentages of cells with aneuploid (mainly hypodiploid) chromosome sets. The increase in hypodiploid set was associated with the loss of C/X group chromosomes by women and loss of G/Y chromosomes by men. On the other hand, an increase in hyperdiploidy related to excessive C group chromosomes was identified in 7 of 19 studies in old women and that related to excessive G/Y chromosomes in 3 of 19 studies in old men.

2.1.2. Relationship between 'Natural' and 'Artifactual' Aneuploidy

Table 2.1.2-1 presents data on changes in diploid numbers of 1,136 karyotypes examined in 40 peripheral lymphocyte cultures of apparently healthy individuals ranging in age from 80 to 114 years; 963 karyotypes of 48 cultures derived from 20 to 48 year-old donors were used as controls (Lezhava and Khmaladze, 1978,1988; Lezhava, 1991, 2001b). The Table shows that aneuploid chromosome patterns occurred in 15.93±1.08 percent of studied karyotypes and 11.31±1.02 percent of the control karyotypes. Of the total karyotype number, hypodiploid cells made 12.94±0.99 (9.76±0.95)[1] percent, of which 6.60±0.74 (4.67±0.68) percent were karyotypes with 45 chromosomes. The incidence of karyotypes with hyperdiploid sets of 47 or more chromosomes at ages from 80 to 114 was 2.99±0.50 (1.55±0.39) percent.

Provided that both germ and somatic aneuploid cells are produced by chromosome nondisjunction during in vitro and in vivo mitotic divisions, a similar incidence of hypodiploid and hyperdiploid cells was expected. However, hypodiploid cell percentages

Control values [1]are Parenthesized

were higher. Cited causes of increased hypodiploidy include a chromosomal lag at the anaphase of human lymphocytes (0.62%, Bochkov, 1971) and an early centromere separation (0.91%, Fitzgerald, 1975); this suggests that most cases of hypodiploidy were associated with a mechanical chromosome loss during preparation making, a phenomenon defined as artifactual variability (Tough et al., 1961; Jacobs et al., 1963). For this reason chromosome counts lower than 45 were interpreted as a secondary event induced by the preparation procedure (Prokofyeva-Belgovskaya and Gindilis, 1965). Mitotic chromosome nondisjunction is the only basis of hyperdiploidy (chromosomes that are lost from other metaphases and mechanically get into cells are readily recognized by spiralization degree) (Table 2.1.2-2).

Our experimental data summarized in Tables 2.1.2-1 and 2.1.2-2 show that numbers of cells with 45 and 47 chromosomes and the total incidence of aneuploidy considerably increase at ages from 80 to 114 years. This observation, supported by most published data, suggests that senescence-specific factors account for the higher incidence of cells with hypo- and hyperdiploid chromosome sets. Relationships between natural and artifactual aneuploidy in old age have not been investigated thus far. However, this problem is of great theoretical and practical importance. The issue of practical importance is whether the quantitative karyotype change is normally associated with aging or represents an abnormal manifestation.

To evaluate the relationships between natural and artifactual aneuploidy, we used the following procedure. Let *PX* denote a distribution of the random variable *X* (a deviant cell chromosome count caused by natural aneuploidy) and let *qy* be a distribution of the random variable *Y* (a number of chromosomes lost because of an artifact)[2]. Since the random variables *X* and *Y* may be considered as independent, the distribution *rz* of the random variable $Z = X - Y$ (a depictor of total aneuploidy) is represented by the formula

$$r_z = \sum_{x-y=z} p_x q_y,$$

(For example, $z = 0$ if the cell has 46 chromosomes).

To evaluate the distribution *PX*, we attempted to follow up its age-dependent change. It proved a difficult task since the *X* and *Y* values could not be elucidated separately, and so our inference of *PX* had to be derived from a sample of the *rz* distribution with the *qy* component unknown.

---

[2] X may be positive (when a chromosome count is excessive) and negative (when chromosomes are missing in the cell). $X = 2$ in a cell with 48 chromosomes and $X = 1$ in a cell with 45 chromosomes. The same applies to *Y*.

Table 2.1.1-1. Age-related incidence of altered chromosome sets in peripheral lymphocyte cultures

No. of subjects	Age (yr)	Sex		No. of meta-phases	Percent of aberrant meta-phases	*Percentage of cells with*			*In old age*				Authors
		male	female			Hypodi-ploidy	Hyper-diploidy	Aneu-ploidy	Increase in Hypodiploidy due to loss of C or G/Y group chromosomes		Increase in Hyperdiploidy due to gain of C or G/Y group chromosomes		
									females	males	females	males	
32	0-14	9	23	940	—	2,9	0,8	3,7					
169	15-64	90	79	6023	—	4,5	1,1	5,6					
46	65+	25	21	1417	—	9,6	2,5	12,1	+	+	—	—	Jacobs et al., 1963
41	0-14	23	18	942	—	5,5	1,1	6,6					
148	15-64	78	70	3644	—	6,0	1,0	7,0	+	+	+	+	Hamerton et al., 1965
5	6-15	+	+	233	—	3,6	0,9	4,5					
10	16-59	+	+	495	—	5,9	2,6	8,5					
5	60+	+	+	236	—	11,9	1,7	13,6	+	+	+	+	Kerkis & Radzhabli, 1966
346	15-74	231	207	12308	—	5,4	0,2	6,6					
92	75+					8,6	0,8	9,4	+	+	+	+	Court Brown, 1966
10	63-91	5	5	564	—	5,7	2,3	8,0	—	—	—	—	Kleisner de Galan, 1966
230	20-59	138	92	9387	—	1,6	0,5	2,1					
99	60+	56	43	8727	—	1,6	0,8	2,4	—	—	—	—	Bloom et al., 1966
38	0-14	23	15	2798	1,4	5,9	1,2	7,1					
113	15-64	61	52	6450	1,7	6,1	1,0	7,1					
20	65+	15	5	1145	2,1	7,6	0,9	8,5	—	—	—	—	Sandberg et al., 1967
20	Newborn	20	—	600	2,2	2,2	0,8	3,0					
26	19-21	26	—	780	3,7	1,7	1,1	2,8					

Table 2. (cont.)

24	67-23	24	—	720	2,0	8,4	2,4	10,8	+		+		Goodman et al., 1969
30	17-35	15	15	3050	1,9	13,4	1,1	14,5					
61	77-93	38	23	4516	4,3	15,7	0,7	16,4	—	+	—	—	Jarvik & Kato, 1970
74	New-born -45	41	33	3694	—	12,0	1,0	13,0					
65	72+	36	29	3373	—	16,2	0,7	16,9	—	—	—	—	Neurat et al., 1970
10	19-39	+	—	361	—	3,6	0,3	3,9					
10	40-59	+	—	408	—	5,1	1,0	6,1					
10	65-87	+	—	734	—	10,1	1,8	11,9	+	—	+	—	Nielsen, 1970
87	30-59	46	41	7973	2,1	4,4	0,9	5,3					
11	60-79	3	8	968	3,6	3,6	0,7	4,3	—	—	—	—	Kadotani et al., 1971
40	New-born	21	19	5117	1,1	—	—	—					
100	20-30	46	54	17800	1,2	—	—	—					
41	60+	17	24	5469	1,4	—	—	—					Bochkov et al., 1972
27	18-58	27	—	6482	2,1	10,4	0,7	11,1					
94	68-98	94	—	9568	4,0	15,2	1,2	16,4	—		+		Jarvik et al., 1974
163	20-39	72	91	17700	0,9	—	—	5,0					
42	40-50	28	14	4300	1,4	—	—	5,7					Sevankaev et al., 1974
22	22-59	22	—	2199	—	5,3	0,3	5,6					
10	60-87	10	—	1000	—	6,4	0,7	7,1	+(X)		+(X)		Fitzgerald, 1975
60	10-13	30	30	1800	6,9	4,2	0,4	4,6					
60	62-96	30	30	1800	11,5	12,0	1,6	13,6	+	—	—	—	Mattevi & Salzano, 1975
17	84-100	11	6	1697	—	15,9	1,6	17,5	+	—			Jarvik et al., 1976
48	0-45	48	—	3594	2,4	—	—	—					
18	60-75	18	—	1459	3,8	—	—	—					Kadotani et al., 1978
48	20-48	23	25	964	1,9	9,8	1,6	11,4					
40	80-114	14	26	1136	4,6	13,9	2,2	16,1	+	+	—	—	Lezhava & Khmaladze, 1978
30	80+	10	20	3850	5,8	18,7	1,0	19,7	+(X)	+	—	—	Isakovich & Vahancik, 1984
228	lwk-90	90	138	1000*	—	8,34	—	8,34	+(X)	+(Y)	—	—	Guttenbach et al., 1995

* 1000 lymphocyte interphase nuclei were analyzed by in situ hybridization for loss of the X and Y chromosomes

Our observations of the aneuploidy incidence allowed us to make several simplifying assumptions. Since metaphases with more than 48 chromosomes were rarely found, it could be assumed that $z3$ was associated with $rz = 0$, i.e. px was 0 with $x3$. Since metaphases, especially round ones, seldom contained less than 42 chromosomes, we could assume that z–5 coexisted with $rz = 0$, which means $qy = 0$ with $y3$, and $px = 0$ with x–3, at least for the round metaphases. Using these assumptions, there was a negligible probability of a loss or excess of 3 or more chromosomes due to natural aneuploidy and a loss of more chromosomes due to an artifact.

The distribution rz could be written as follows

$$r_2 = p2q0$$

$$r_1 = p2q1 + p_1q_0$$

$$r_0 = p2q2 + p_1q_1 + p_0q_0$$

$$r_{-1} = p_1q_2 + p_0q_1 + p_{-1}q_0$$

$$r_{-2} = p_0q_2 + p_{-1}q_1 + p_{-2}q_0$$

$$r_{-3} = p_{-1}q_2 + p_{-2}q_1$$

$$r_{-4} = p_{-2}q_2 \quad (1)$$

By using a usual two-sample $X2$ test instead of the previous formulas it could be easily found (Table 2.1.2-2) that the distribution rz was similar in men and women aged 20 to 48 years.

In contrast, the distribution rz significantly varied in elderly men and women. Moreover, this distribution varied in the control group, which was found by calculating the statistics:

$$T_z = \left(n_{1z} - n_z \frac{N_1}{N_1 + N_2} \right) \left(\frac{n_z N_1 N_2}{(N_1 + N_2)^2} \right)$$

where $N1$ and $N2$ were numbers of cells compared in the study and control groups, and $n1z$ and $n2z$ were frequencies of cells with 46 + z chromosomes; $nz = n1z + n2z$.

Thus, when we tested the hypothesis that 1 – $r0$(s.m.) = 1 – $r0$ (c.m.), where s.m. stands for senile men and c.m. for control men, the T statistics proved similar (see Table 2.1.2-1):

$$T = (117 - 1850.56)(185\ 0.560.44)-1/2 = 2.0.$$

This value was above the 2.3 percent point of a normal standard distribution, suggesting a higher probability of aneuploidy in senile men as compared to the control group. According to Table 2.1.2-1, the T value in senile women also differed from controls:

$T = (17 - 230.52)\ (230.52\ 0.48) - 1/2 = 2.10$

which was above the 1.8 percent point of a normal distribution. This means that a probability of cells with 47 chromosomes was significantly higher in the study group of senile women as compared to the control group.

However, evidence that the distribution *rz changes* is not sufficient to conclude that its component, the distribution *PX*, changes too. In particular, it needs to be ascertained whether or not changes in *rz* are attributable to changes in artifactual aneuploidy distribution only. The following simple test served this purpose.

Assuming that artifactual chromosome loss in metaphases of senile men was more predictable than in the control group and distributions *px* of natural aneuploidy were similar, we would have obtained from (1):

$$r+(\text{s.m.}) - r+(\text{c.m.}) < 0$$

$$r\text{-}(\text{s.m.}) - r\text{-}(\text{c.m.}) > 0 \qquad (2)$$

where $r+ = r1 + r2$ were a probability of hyperdiploidy and $r- = r{-}1 + r{-}2 + r{-}3 + r{-}4$ were a probability of hypodiploidy in the respective groups.

On the contrary, if fewer chromosomes were lost due to artifact in the study group of senile men as compared to the control group, the result would have been:

$$r+\ (\text{s.m.}) - r+(\text{c.m.}) > 0$$

$$r\text{-}\ (\text{s.m.}) - r\text{-}(\text{c.m.}) < 0. \qquad (3)$$

The statistics calculated in accordance with Table 2.1.2-2 were:

$T+ = (16 - 240.56)\ (240.560.44) - 1/2 = 1.01$

$T- = (101 - 1610.56)\ (1610.560.44) - 1/2 = 1.72$

where 16 was the number of hyperdiploid metaphases and 101 the number of hypodiploid metaphases in senile men. The data (1.01 was below the upper 15 percent point of a normal distribution while 1.72 was above the upper 5 percent point) suggested that it was the first set of inequalities that was valid, i.e. the probability of artifact was higher in the study group of senile men than in the control group. It was still unclear whether the distribution *px* (s.m.) differed from *px* (c.m.).

Table 2.1.2-1. Distribution of chromosome counts in lymphocyte cultures from subjects at 80 years and older

No. of Subjects	Sex	Age (yr)	No. of karyotypes	2n 42 (4)	43 (3)	44 (2)	45 (1)	46 (0)	47 (1)	8 (2)	(%) Hypodiploidy	Hyperdiploidy	Aneuploidy
14	F	80-108	407	8	4	15	19	343	17	1	11,30±1,56	4,42±1,01	15,72±1,80
26	M	80-114	729	7	13	25	56	612	15	1	13,85±1,27	2,19±0,54	16,05±1,35
40	F,M	80-114	1136	15	17	40	75	955	32	2	12,94±0,99	2,99±0,50	15,93±1,08
23	F	20-46	390	4	5	8	17	349	6	1	8,71±1,42	1,79±1,29	10,51±1,55
25	M	24-48	573	6	10	16	28	505	8	0	10,47±1,27	1,39±0,48	11,86±1,35
48	F,M	20-48	963	10	15	24	45	854	14	1	9,76±0,95	1,55±0,39	11,31±1,02

Table 2.1.2-2. Distribution of missing and additional chromosomes in karyotypes from subjects at 80 years and older

Age (yr)	Sex	No. of karyotypes	No. of chromosomes	Chromosome groups								Total No. of aneuploidy
				A	B	C	D	E	F	G	Y	
80-108	F	407	42	4	5	9	4	3	4	3	—	8
			43	1	1	5	2	1	1	1	—	4
			44	1	3	9	6	5	2	4	—	15
			45	—	1	10	2	1	2	3	—	19
			47	—	—	6	2	1	2	6	—	17
			48	—	—	1	—	—	—	1	—	1
80-114	M	729	42	2	3	2	4	4	3	1	2	7
			43	2	1	7	6	6	1	10	3	13
			44	—	4	8	1	11	5	6	11	25
			45	2	6	9	5	5	4	10	14	56
			47	—	—	3	3	3	1	8	—	15
			48	—	—	—	—	—	1	1	—	1
Total 80-114	F, M	1136		12	24	79	7	40	26	54	30	181
20-46	F	390	42	1	1	4	3	2	2	3	—	4
			43	—	—	3	2	2	3	5	—	5
			44	—	1	10	1	1	2	1	—	8
			45	—	—	6	2	2	5	2	—	17
			47	—	—	2	—	1	1	2	—	6
			48	—	—	2	—	—	—	—	—	1
20-48	M	573	42	—	1	4	5	5	3	4	2	6
			43	1	1	1	3	3	4	5	2	10
			44	1	1	4	3	1	3	5	4	16
			45	—	1	7	3	3	1	5	8	28
			47	—	—	4	—	2	—	2	—	8
			48	—	—	—	—	—	—	—	—	0
Total 20-48	F, M	963		3	5	67	14	17	24	34	16	109

A similar analysis of data for senile women (Table 2.1.2 -2) resulted in $T+ = 2.10$ and $T- = -1.40$. The probability of natural aneuploidy was obviously higher in senile women as compared to controls. Our results indicated that the probability of natural aneuploidy was substantially greater in senile women and that of artifactual aneuploidy in senile men. The problem of natural aneuploidy in men remained unclear.

Distribution patterns of missing and supernumerary chromosomes in karyotypes of individuals aged 20 to 48 years and 80 to 114 years (Lezhava, 1991,2001a) have shown that the rates of chromosome loss in elderly men and women were significantly above those in control groups. Thus, the incidence of group A and B chromosome loss was higher in elderly women (Table 2.1.2.-2); an increased chromosome loss in karyotypes with 45 chromosomes was observed only in the C group (Table 2.1.2 -2).

A greater incidence of supernumerary chromosomes in elderly women was characteristic for the G group (Table 2.1.2.-2). As regards nuclear chromosome loss in elderly men, it was statistically significant for the B group and nearly significant for the A, E groups and Y chromosomes.

Table 2.1.2-2 presents evidence for an increased incidence of karyotypes with supernumerary chromosomes in old men. Most of hyperdiploidy was related to G group chromosomes.

2.1.3. Premature Disjunction of Centromeres

Premature centromere disjunction in mitotic sister chromatids produces cells with 45 and 47 chromosomes. The abnormal chromatid disjunction (aneuploidy) is underlied by a lesion in the mitotic achromatic system (Vig, 1983; Nath et al., 1995).

Metaphases with chromosome spiralization indices of 29 to 31 percent were assessed in order to delineate the occurrence of cells with prematurely separated centromeres in old individuals (Gindilis, 1966). It was found to be comparable in lymphocyte cultures of individuals aged 80 years or older and controls; only C group chromosomes with premature centromere disjunction were more common in old women as compared to women aged 20 to 48 years (Fig. 2.1.3-1).

This evidence provides support to the opinion that the incidence of cells with premature centromere disjunction of C group chromosomes increases with age in women (Fitzgerald, 1975; Nakagome et al., 1984; Mukherjee and Weinstein, 1988; Griffin, 1996). Mukherjee and Weinstein (1988) indicates that changes in the sequence of centromeric separation of certain chromosomes do occur in lymphocytes during aging in humans.

2.1.4. Polyploidy (Endoreduplication)

A major mechanism of polyploid somatic cell production in animals and humans is endomitosis, or chromosome set duplication in the enveloped nucleus without a subsequent mitosis (Geitler, 1939). During the 'endomitotic cycle', chromosomes reduplicate at the interphase without leaving it (Painter and Reindorp, 1939) and undergo spiralization and despiralization, as mitotic chromosomes (Levan and Hauschka, 1953; Prokofyeva-Belgovskaya, 1968). Endomitosis was originally described in Gerris lateralis (Geitler, 1939), and its various modifications were subsequently documented in numerous organisms (Levan and Hauschka, 1953).

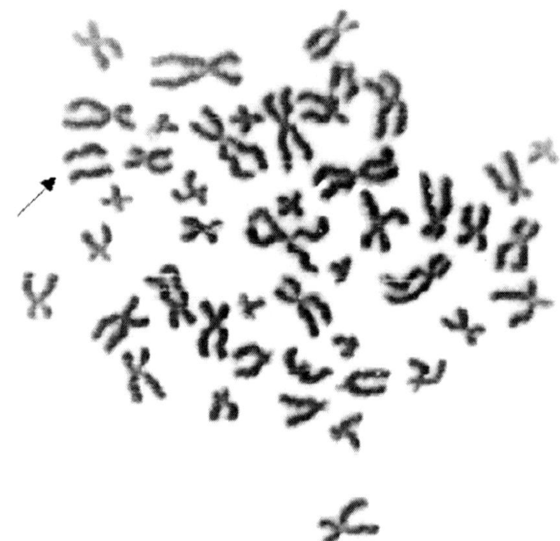

Fig. 2.1.3-1. Metaphase of an 80-year-old woman. Premature chromatid disjunction in C group chromosome

Cinematographic studies of a HeLa strain by Hsu and Moorhead (1956) gave a classification of polyploidy in relation to mitotic stages and a description of its mechanisms:

(1) Intrareduplication, or endoreduplication, described by Levan and Hauschka (1953). The polyploidization process occurs at the interphase and results in diplosomes, triplosomes or more multiple chromosomes.
(2) Proreduplication (most of known endomitoses are represented by this type). Spindle abnormality arises at the prophase. Any events inducing this version of polyploidy occur when the nuclear membrane is available.
(3) Metareduplication. It is associated with the C mitosis which begins at the metaphase. Spindle abnormality at this phase results in a random cytoplasmic distribution of chromosomes. Cumulative evidence indicates that the incidence of metareduplication in cultured lymphocytes from middle-aged individuals is within the range of 0.3-0.5 percent, but old age increases it to 0.7 percent (Goodman et al., 1969; Kadotani et al., 1978; Lezhava, 1968, 2001b).
(4) Anareduplication. Spindle abnormality at the anaphase produces chromosome clustering in the equatorial plane.
(5) Teloreduplication. A simple fusion of daughter cells during the cytokinesis or after it may result in multinuclear cells.

Endoreduplication, which was first studied in murine ascites tumors by Levan and Hauschka (1953), represents one of endomitosis forms. These investigators suggested that endoreduplication is associated with microscopically invisible intranuclear chromosome duplication during the interphase, after which no cell divisions occur.

Most of the short-term peripheral lymphocyte cultures of normal subjects did not display endoreduplicated chromosomes (Moorhead et al., 1960). The incidence of endoreduplicated chromosomes in long-term fibroblast cultures varied from 0.5 to 5.8 percent. Intraindividual variability in the incidence of nuclei with endoreduplicated chromosomes was found to be greater than interindividual ones (Schwartzacher and Schnedl, 1965). A study in 70 spleen cultures also demonstrated a high incidence (1 to 8 percent) of such nuclei (Bain and Gauld, 1964).

Our overview of reported findings suggests that short-term culturing (up to 4 days) of lymphocytes from normal middle-aged persons typically revealed no cells with endoreduplicated chromosomes or only a small proportion (0.7 percent) of such cells. A significantly greater number of cells with endoreduplicated chromosomes has been identified by longer culturing (Dvalishvili, 1989). In some cases the incidence of endoreduplicated chromosomes increases with age in short-term blood cell cultures (Kadotani et al., 1978; El Zawahry and Khishin, 1980). Karyotipic and phenotypic changes were found in human adult endothelial cells. An increase in the poliploid cells was also found near the end of the in vitro life span in 6 cell lines (Johnson et al., 1992).

The incidence of polyploid chromosome patterns has been evaluated (Lezhava, 1968, 2001a) in blood cultures of 20- to 114-year-old individuals using the Hsu and Moorhead (1956) classification of endomitosis. The data of Table 2.1.4.-1 show that in a total of 70 cultures of old subjects, cells with chromosome metaredublication made up 0.55 ± 0.15 percent (women, 0.65 ±0.25 percent; men, 0.48 ±0.19 percent). These values were within the control range (Table 2.1.4-1.)

Table 2.1.4–1. Incidence of polyploidy in peripheral lymphocyte cultures at ages of 20 to 114 years

No. of subjects		Sex	Age (yr)	No. of cells	Cells with metareduplication No.	Cells with metareduplication %	Cells with endoreduplication No.	Cells with endoreduplication %
	30	F	80-108	1064	7	0.65±0.25	4	0.37±0.19
	40	M	80-114	1262	6	0.48±0.19	8	0.63±0.22
Total	70	F,M	80-114	2326	13	0.55±0.15	12	0.51±0.15
	23	F	20-46	552	2	0.36±0.25	—	—
	25	M	24-48	573	1	0.18±0.18	1	0.18±0.18
Total	48	F,M	20-48	1125	3	0.26±0.15	1	0.08±0.08

Twelve of 70 lymphocyte cultures of individuals aged 80 years or older exhibited 0.51±0.15 percent of cells with endoreduplicated chromosomes (diplochromosomes) (Fig. 2.1.4-1.); 0.37±0.19 percent of the cells occurred in 4 female cultures and 0.63±0.22 percent in 8 male cultures. A statistical analysis demonstrated a significant difference of diplochromosome incidence in elderly and control individuals.

Diplochromosomes and diploid cell chromosomes are similar in their morphology, satellite stalks of acrocentric chromosomes and the ability of acrocentrics to form associations.

Therefore, the evaluation of peripheral lymphocyte cultures for cells with polyploid chromosome sets showed a higher incidence of metaphase diplochromosomes at 80 years or older ages, as compared to middle age. Percentages of cells with metareduplicated chromosomes were similar at ages of 20 to 48 years and 80 to 108 years.

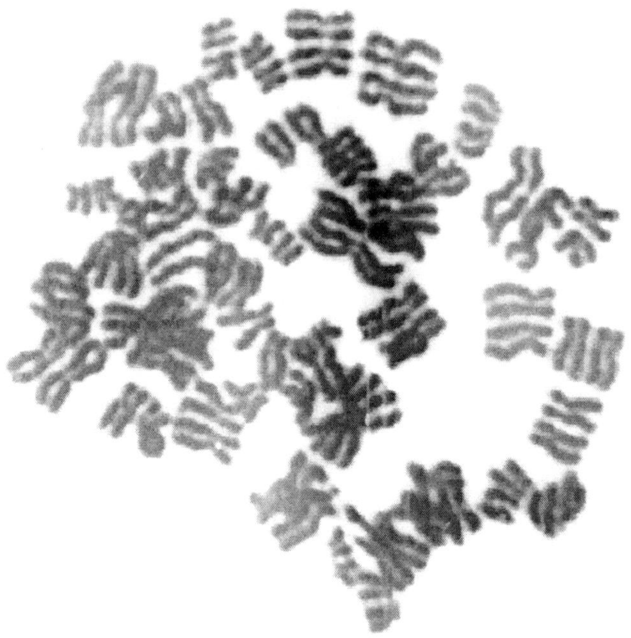

Fig. 2.1.4 -1. Metaphase with chromosomes endoreduplication.

2.2. Aberrations

2.2.1. Incidence of Aberrant Cell and Chromosome Aberrations

Numerous cytogenetic studies have indicated that the incidence of chromosome aberrations in blood cells increases with advancing age. Initial evidence for a greater incidence of cells with chromosome aberrations in 90- to 95-year-old individuals was reported by Bochkov et al. (1968). That study examined spontaneous chromosome aberration rates in 75 first-mitosis peripheral lymphocyte cultures of 59 males and females of different ages. A major proportion of chromosome aberrations was found in cells with paired (49 percent of total aberrations) and single fragments. A mean incidence of chromosome aberrations for all age groups was 1.6 percent; the mean was 1.8 percent for newborns, 0.6 percent at 10-11 years, 1.6 percent at 20-29 years, 1.6 percent at 30-39 years, 0.8 percent at 50-70 years and 3.1 percent at 90-95 years. Variability of spontaneous chromosome aberration rates was statistically nonsignificant. Follow-up evaluations showed a comparable intra- and interindividual variability of the rates.

Bochkov et al.(1972) in a later study, found no correlation between the incidence of cells with chromosome aberrations, age and first- or second-mitosis cultures which evaluated spontaneous aberration rates in 28,386 cells of 228 blood cultures derived from 181 subjects. Sandberg et al. (1967) tested 10,393 blood cells from 171 subjects for chromosome aberrations; the study involved 3 groups aged: below 14 years, 25 to 64 years and 65 years or more. Age had no effects on percentages of cells with structural chromosome aberrations (Table 2.1.1-1). Astonishingly, Goodman et al. (1969) found a higher incidence of chromosome aberrations in lymphocytes from 19- to 21-year-old subjects, as compared to ages of 67 to 93 years. Most of the studies reported an increase in cells with chromosome aberrations and the incidence of aberrant chromosomes with age.

A study in 61 twins aged 77 to 93 years (23 men, 38 women) indicated a higher proportion (4.27 percent) of metaphases with aberrant chromosomes in elderly women as compared to controls (Jarvik and Kato, 1970).

Kadotani et al. (1971) examined blood cultures of 46 subjects from birth to 45 years of age and 18 individuals of 60 to 75 years and found that the incidence of cells with chromosome aberrations increased with age (Table 2.1.1-1).

In the study of Jarvik et al. (1974), lymphocyte cultures were assayed in 121 subjects in the age ranges from 18 to 58 years and 68 to 98 years; percentages of cells with chromosome aberrations appreciably rose with age. The most common aberrations were chromatid and isochromatid breaks.

A still higher incidence of chromatid and chromosome aberrations (11.5 percent) in senile individuals has been reported by Mattevi and Salzano (1975a) (Table 2.1.1-1). According to Kuznetsova and Zaritskaia (1986), longevity is associated with higher rates of spontaneous chromosome aberrations. The incidence of lymphocytes with aberrations was found to be 6 percent in old and 1 percent in young individuals. The aberrations typically occurred as single and paired fragments and were most common in B group chromosomes. An accumulation of mutations on their own or together with other age-related changes may contribute to aging and the development of age-related pathologies (King et al., 1994). The aim of this investigation was to assess the extent of DNA mutations as a function of age in humans. The mutant frequency (MF) at the hypoxanthine-guanine phosphoribosyl-transferase (HGPRT) locus was assessed in lymphocytes isolated from male volunteers in each of three age groups (35-39, 50-54 and 65-69 years). Results show that the mean MF in the 65-69 years group was approximately twice that in the 35-39 and 50-54 years groups (4.1/10(6) cells, 1.9/10(6) cells and 1.79/10(6) cells, respectively) increasing by about 1.33% per year, after 54 years. In addition, there was an increased frequency of chromosomal aberrations in the 65-69 years group compared to the other two age groups.

Lymphocytes from normal donors have been evaluated for chromosome aberrations in the study of Prieur et al. (1988). An analysis of about 1,000 metaphases from 2 newborns, 4 young (25 to 36 years) and 2 old (73 and 77 years) subjects showed that the frequency of aberrations in the bands 7p14, 7q35, 14q11, 14q12 and 14q ter was 0.0043 in the newborns and 0.0024 in the adults. Rearrangements in other bands were more common in the elderly (0.038) than in the young adults (0.025) and newborns (0.013). Damages of the first type were thought to originate with immunoglobulin genes or related genes and those of other types to be a function of aging. Spontaneous baseline frequencies of chromosome aberration, micronucleus counts and cell division were analysed in peripheral lymphocytes of 127 normal healthy individuals in vitro (Ganguli, 1993). 100 metaphases were observed for chromosome aberrations and 1,000 cells each for micronucleus counts and mitotic index. Regression analyses were carried out to see the effect of age on spontaneous abnormalities. The correlation of aberrations, micronucleus formation and mitotic index with donor's age is highly significant. The elevation of abnormalities and depression of mitotic index were linear with the increase of donor's age, with a higher frequency in males. Aged males and females from the age range of 40-70 years showed larger numbers of aberrations. The FISH technique with whole chromosome probes ('chromosome painting') provides an efficient approach for detecting structural chromosome aberrations in human lymphocytes (Ramsey, 1995; Vorobtsova, 2001). This rapid and sensitive technique is an effective tool for quantifying chronic exposure to environmental agents which may result in an accumulation of cytogenetic damage with age. Ramsey applied this technology to a normal, putatively unexposed, population to document the relationship between age and the accumulation of cytogenetic damage, as well as to establish a baseline frequency of stable aberrations. Using probes for chromosomes 1, 2 and 4 simultaneously, the equivalent of 1,000 metaphases was scored for stable and unstable aberrations from each of 91 subjects ranging in age from newborns (umbilical cord blood; $n = 14$) to adults aged 19 to 79 years. Each subject (or one parent of each newborn) completed an extensive questionnaire to identify possible lifestyle factors that may have influenced the frequency of cytogenetic damage. These findings show a significant increase in stable aberrations (translocations and insertions) with age ($p<0.001$). Ramsey also observed age-related increases with dicentrics ($p<0.001$) and acentric fragments ($p<0.001$). Relative to the frequencies observed in cord blood, the frequencies of stable aberrations, dicentrics, and acentric fragments in adults aged 50 and over were elevated 10.6-fold, 3.3-fold and 2.9-fold, respectively.

Evidence on the incidence and patterns of chromatid gaps is an extension of knowledge of structural chromosome abnormalities. The chromatid gaps indicate a continuous metaphase chromatid; its distal portion is not lost at the anaphase, and fragments are not produced (Evans, 1963; Brogger, 1975). The gaps are thought to be induced by damage to chromatid proteins that have critical roles in chromosome thread packing (Brogger, 1975).

The chromatid gaps as structural chromosome abnormalities other than true deletions were first recognized by Revell (1955). Evidence to support it came from several studies (Evans, 1963; Conger, 1967; Brogger, 1971). The incidence of the gaps in lymphocytes cultured from individuals of middle age varies from 4.51 to 7.31 percent (Brogger, 1971; Nordenson et al., 1978). Gap patterns in metaphase chromosomes are not random. Nordenson et al. (1978) have reported a correlation between the incidence of chromosome aberrations and chromatid gaps. These authors suggested that the gaps may be used as sensitive indicators of mutagenic exposures.

Therefore, the majority of studies have yielded a strong relationship between advancing age and the incidence of cells with chromosome aberrations and aberrant chromosomes.

Fig. 2.2.1-1 presents evidence on the incidence of cells with aberrant chromosomes and structural chromosome aberrations (Lezhava and Khmaladze, 1978; Lezhava, 1999, 2001a, b). This study screened 2,326 metaphases from 70 subjects (30 women, 40 men) ranging in age from 80 to 114 years. A mean percentage of aberrant cells and a mean number of chromosome aberrations in women and men aged 80 years or older (respectively 4.08 ± 0.41 percent and 5.15 ± 0.45) were significantly higher than control values (respectively 1.42 ± 0.35 percent and 2.04 ± 1.02).

The incidence of cells with chromosome aberrations in 80- to 90-year-old subjects was 4.75 ± 0.71 percent for 25 women and 3.06 ± 0.54 percent for 31 men; these means were also above those of 20- to 48-year-old individuals. The incidence of aberrant cells in men aged 91 to 114 years (5.62 ± 1.45 percent) was higher than that in women aged 91 to 108 years and control subjects.

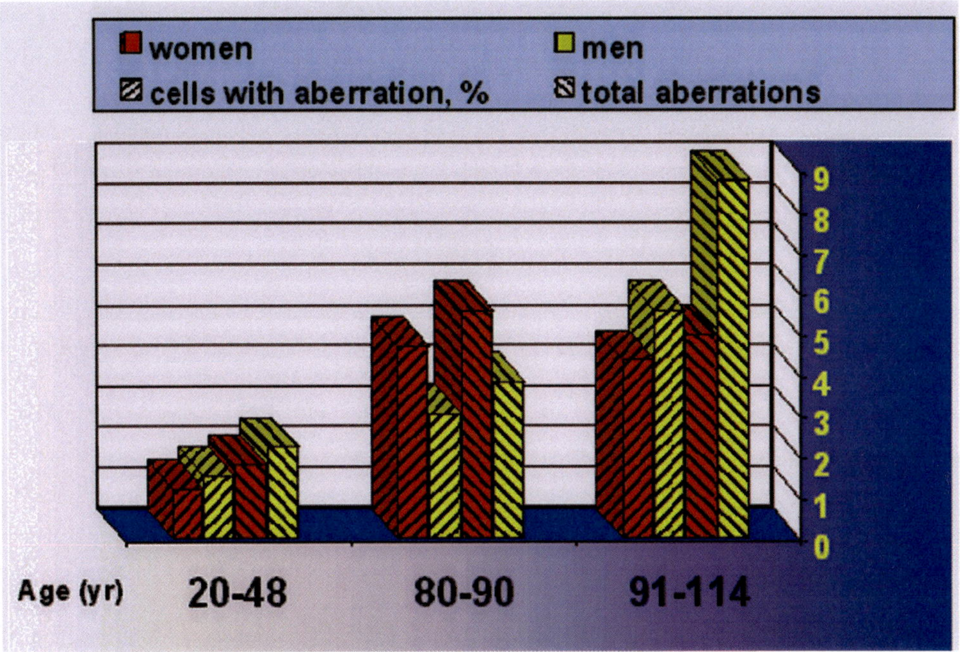

Fig. 2.2.1-1. Spontaneously structural chromosome aberration at 80 years and over

The analysis of aberrant chromosomes in senile subgroups demonstrated similar sex-related patterns; a mean number of such chromosomes per 100 cells was 5.15 ± 0.45, which was much higher than in 20- to 48-year-old subjects.

Structural chromosome aberrations were identified in 95 metaphases (4.08 ± 0.41 percent). Eighty-one metaphases showed a lesion in only one chromosome (3.48 ± 0.38 percent), 8 in two (0.34 ± 0.12 percent), 3 in three (0.12 ± 0.07 percent), 2 in four (0.08 ± 0.08 percent) and 1 in seven (0.04 ± 0.04 percent) chromosomes (Fig. 2.2.1 -2.).

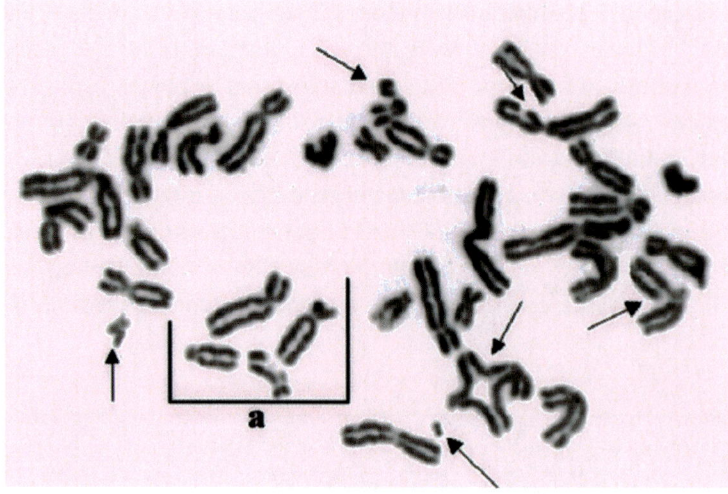

Fig. 2.2.1-2. 114-year-old man's metaphase with aberrant chromosomes. a - association of telomerie regions.

2.2.2. Types of Chromosome Aberrations

The analysis of aberration types showed that aberrant metaphases occurred in the elderly as aberrations of chromosome and chromatid types. Common chromosome damages were single and paired fragments; dicentrics, translocations, chromatid exchanges and atypical chromosomes were rare. Numbers of the chromatid-type aberrations (chromatid and isochromatid fragments, chromatid exchanges) in the senile and control populations were greater compared to the chromosome-type (paired fragments and chromosome exchanges). At ages over 80, the mean number was 1.31 ± 0.22.

Control (20- to 48-year-old) and old subjects exhibited a significant difference in the incidence of chromatid and chromosome aberrations; single and paired fragments, chromatid and chromosome exchanges were far more common at 80 and older ages. It should be noted that a relative incidence of the chromatid- and chromosome-type aberrations was little affected by advancing age.

The distribution of breaks and gaps was evaluated in relation to chromosome groups and sex in subjects aged 80 years and more. Empirical incidence estimates were obtained from lengths of metaphase chromosomes.

Group-related chromosome tear patterns at ages from 80 to 114 years were found to be nonrandom and uneven. The incidence of B and C group chromosome damage in senile women was significantly above a theoretical range, while group A1, A2, B and Y chromosomes were commonly affected in men. About 92 percent of the breaks were located in long arms. Chromosome gaps were much rarer at 80 and older ages. The mean gap incidence was 4.74 ± 0.90 percent in senile women and 5.15 ± 0.96 percent in men. Gap distribution in chromosome groups proved uneven and almost random.

2.2.3. Chromosome Aberrations Induced by Heavy Metal Salts

Recently, much attention has been given to the effects of heavy metal ions on different biological processes. This focus has been prompted by roles they play at all stages of vital functions of organisms. The metal ions penetrate into the cells as cations. The levels of ion absorption and concentration are controlled by the special protein systems providing their transportation to the corresponding place, and inclusion into the required enzymes system as well (Sharma amd Talukder, 1987;Ercal et al.,2001; Kawanishi et al.,2002).

In addition, the participation in normal biological processes (transformation of groups, redox and hydrolysis processes, physiological control and trigger mechanisms, etc.), the metal ions can display toxic action and many of them may have mutagenic effect as well (Clayson, 1994; Kawanishi et al., 2002). The mutagenic features of heavy metal salts, in turn, to a great degree depended on different factors. In particular, it may be highly increased in a model system of aging in plant cells characterized by high spontaneous mutation frequency (Jokhadze, 1977; Maeng et al., 2004).

The redox-active metals, such as copper, iron, undergo redox cycling, catalyzing ROS generation from various organic carcinogenes, resulting in oxidative DNA damage, whereas redox-inactive metals, such lead, cadmium and others deplete cell major antioxidants, particularly thiol-containing antioxidants and enzymes(Kawanishi et al.,2002). Catalytic copper is believed to play a central role in the ROS formation, such as O2 - and OH radicals, that bind very fast to DNA and produce damage by breaking the DNA strands or modifying the bases and/or deoxyribose leading to carcinogenesis (Theophanides and Anastassopoulou, 2002).

The increased spontaneous mutation level was observed in cells *in vivo* (Lezhava, 1984b) and *in vitro* aging (Ronne, 1980; Dvalishvili, 1989): the chromosome aberration induced by the inorganic salts of copper and cadmium in human lymphocyte cultures were studied during *in vivo* and *in vitro* aging.

Lymphocytes were cultured without adding serum and antibiotics (Jokhadze and Lezhava, 1994). Lymphocytes from two age groups were cultivated: 20-25 years old (young subject) and 80-93 years old (old subjects). Cells from young subjects were cultivated for 50 hr (control) and 144 hr (*in vitro* aging) where as cells from old subjects were cultivated for 50 hr only. The same concentrations (0.5 x 10-4M) of copper and cadmium salts were present throughout the 50 hr cultures and were added at the 96 hr in the case of 144 hr cultures.

The analysis of data obtained at 50 hr shows that copper and cadium ions caused a dramatic increase in the number of cells with chromosome aberrations. The frequency of aberrant cells in the presence of copper sulfate was 6.5 ± 1.23% and 5.45 ± 0.97% in the presence of cadmium chloride (versus 1.72 ± 0.71% in control) (Table 2.2.3-1). An increasing number of cells with spontaneous chromosome aberrations was observed in 144 hr cultures (*in vitro* aging)—5.25 ± 1.11%, when compared to 50 hr. This is in accordance with the published data (Dvalishvili, 1989). The presence of copper sulfate in culture medium increased the value to 12.20±1.62%. Cadmium chloride did not trigger any increase in the 144 hr cultures (4.04±0.98%) (Table 2.2.3-1).

Table. 2.2.3-1. Chromosome aberrations induced by heavy metals in human lymphocytes

Type of heavy metals	50 hr-cultures from subjects aged 20-25yr (control)		144 hr-cultures from subjects aged 20-25yr (in vitro aging)	
	No. of metaphases	Percent of aberrant metaphases	No. of metaphases	Percent of aberrant metaphases
Control	640	1.72±0.51	400	5.25±1.11
$CuSO_4$	400	6.50±1.23	410	12.20±1.62
$CdCl_2$	550	5.45±097	396	4.04±0.98

Similar results with respect to the induction of chromosome aberrations with these salts were obtained in 50 hr lymphocyte cultures of subjects aged 80 and over (Fig. 2.2.3-1). Copper ions caused a considerable increase in cell number with aberrant **cFi** chromosomes aged subjects (14.25 ± 1.74% in comparison with 3.94 ± 1.08% in control) (Jokhadze and Lezhava, 1994), while cadmium did not raise the level of cells with changes of chromosomes.

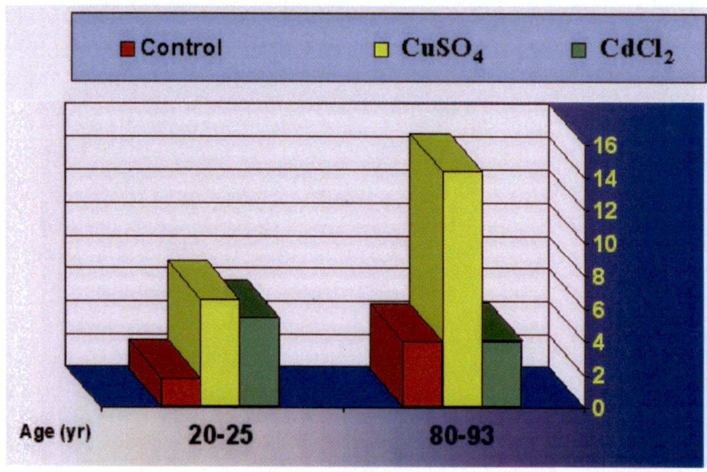

Fig. 2.2.3-1. Chromosome aberrations induced by heavy metals in human 50 hr lymphocytes cultures.

Discussion

Aneuploidy

An increase in the human aneuploid lymphocyte number with age, first reported by Jacobs et al. (1961), was confirmed by multiple studies (Jacobs et al., 1963; Hamerton et al., 1965; Kerkis and Radzhabli, 1966; Court Brown et al, 1966; Kerkis et al., 1967; Sandberg et al., 1967; Neurat et al., 1970; Nielsen, 1970; Kuznetsova, 1972; Jarvik et al., 1974; Sevankaev et al., 1974; Mattevi and Salzano, 1975a; Izakovic and Vahancik, 1984; Abruzzo et. al., 1985; Lezhava and Khmaladze, 1988; Mukherjee and Wallace, 1990; Hando et al.,

1994; Catalan et al., 1995; Nath et al., 1995; Robbins et al., 1995; Griffin, 1996; Lezhava, 1999, 2001a; Warburton, 2005). Evidence from these studies agrees with our findings in metaphases from individuals aged 80 years or older: a mean incidence of aneuploid cells was 11.31±1.02 percent at 20 to 48 years and 15.93±1.08 percent at 80 and older ages.

Controversy arose in some studies that reported a comparable incidence of aneuploid cells in young and old age (Kleisner de Galan, 1966; Bloom et al., 1967; Kadotani et al., 1971; El Zawahry and Khishin, 1980). Plausible causes of the variability of results, as listed by Jarvik et al. (1974), are geography of the studies (Japan, Europe, etc.), mechanical factors (culture methodologies, preparations), and analytical bias, endogenous and exogenous impacts (radiation, viral infection, drugs, chemicals, nutrition).

Hypodiploidization was almost uniformly indicated as a basis of higher aneuploidy in old age. Furthermore, Kerkis et al. (1967) reported that lymphocyte resistance is lowered by aging: this may induce artifactual hypodiploidy when preparations are made. In this case chromosomes are most likely to be lost from the periphery of the nucleus. Since this region is a major location of late replicating chromosomes X and 2, 4-5, 13, 18, 21-22 (one pair) (Miles, 1961; Miller et al., 1963a, b; Prokofyeva-Belgovskaya, 1986; Ockey, 1986, Chandra et al., 1972), it was predictable that increasingly more metaphases would lose these chromosomes with aging. Actual evidence was obtained for loss of the X and Y chromosomes (Jacobs et al., 1963; Hamerton et al., 1965; Kerkis and Radzhabli, 1966; Nielsen, 1970; Demoise and Conard, 1972; Fang et al., 1975; Fitzgerald, 1975; Mattevi and Salzano, 1975a; Richard et al., 1994; Catalan, 1995 and others), the 21-22-Y chromosome (Jacobs et al., 1961; Neurat et al., 1970; Jarvik and Kato, 1970; Nath et al., 1995; Robbins et al., 1995, and others) and chromosomes of group E (Zang and Zankl, 1970; Buckvic et al.,2001). A higher loss of B group chromosomes has been documented by us in subjects from 80 to 114 years of age, a confirmation of the findings of Kuznetsova (1972).

Furthermore, we subscribe to the opinion of Kerkis et al. (1967) that an aging-associated decline in leukocyte resistance is part of generalized changes in cells of senescent organisms, occurring as an increase in cytoplasmic viscosity and secondary abnormality in chromosome disjunction. It is of noteworthy that with increasing age somatic cells display a higher incidence of mitotic chromosomes with premature centromere separation (Fitzgerald, 1975; Mukherjee and Weinstein, 1988; Griffin, 1996), inactivation or loss of kinetochores (14 women from 66 to 87 years of age demonstrated their higher incidence of 0.087 as compared to 0.003 in younger women) (Nakagome and Abe, 1982; Nakagome et al., 1984), abnormalities in mitotic chromosome segregation (Uchida et al., 1975; Bajnoczky and Mehes, 1985) and age-dependant higher rates of maternal chromosome nondisjunction associated with infantile disease (Penrose, 1962; Akesson et al., 1966; Lange et al., 1975; Hassold and Chiu, 1985). These findings suggest that age- and sex-related endogenous factors may induce some chromosome loss in a senile organism (Neurat et al., 1970). The overall implication is that aging increases a cell population with abnormal chromosome sets.

However, exact patterns of aging-related natural and artifactual aneuploidy remain unknown. Our studies have demonstrated a higher probability of artifactual aneuploidy in elderly men as compared to control populations. Supportive evidence is offered by studies of Nielsen (1970) and Mattevi and Salzano (1975a). The latter study presented an extensive survey (Jacobs et al., 1963; Hamerton et al., 1965; Bloom et al., 1967; Sandberg et al., 1967;

Cadotte and Fraser, 1970; Jarvik and Kato, 1970; Neurat et al., 1970; Nielsen, 1970; Kadotani et al., 1971). The higher rates of artifactual aneuploidy in elderly men may be related to endogenous determinants of chromosome loss (E and G, one pair; Y), which depends on chromosome sizes and nuclear locations. It is uncertain from these studies whether the incidence of natural aneuploidy is increased in elderly men, although a high probability of that was suggested by Jacobs et al. (1963).

In our study, the probability of artifactual aneuploidy in senile women was found to be lower ($T- = -1.4$) and that of natural aneuploidy significantly higher ($T+ = 2.1$). The original study of Mattevi and Salzano (1975a) has examined these probabilities in women aged 64 to 96 years and girls aged 10 to 13 years to yield $T+ = 5.6$ and $T- = -3.6$. this implies a significantly higher rate of natural aneuploidy in old women. However, cumulative evidence contradicts this, and the discrepant results are difficult to interpret. It appears that selective impairment of chromosome disjunction during the aging process (Fitzgerald, 1975; Uchida et al., 1975; Bajnoczky and Mehes, 1985, 1988; Soewarto et al., 1995) accounts for a greater occurrence of natural aneuploidy in cells of old women.

Polyploidy

Numerous studies of somatic cell chromosomes in animals and humans suggest that percentages of polyploid cells (with chromosome metareduplication) in some parenchymatous organs increase with age (Lezhava, 1968; Goodman et al., 1969; Kadotani et al., 1978; Raes et al., 1984). In addition, the incidence of metaphases with diplochromosomes has been found to be increased by aging in humans (Kadotani et al. 1978; Lezhava, 1991). Confirmatory evidence for the higher incidence of polyploid cells with endoreduplicated chromosomes in males and females older than 80 has been provided by Goodman et al. (1969), Lezhava (1968), Kadotani et al., (1978), El Zawahry and Khishin (1980). Explanations for the increasing polyploidy may lie in hormonal changes of aging (Goodman et al., 1969; Walter et al., 1981; Lezhava, 1999, 2001 a,b).

Chromosome Aberrations

Spontaneous chromosome aberrations in human somatic cells came into sharp focus with recognition of important modifiers of induced mutagenesis rates. Spontaneous aberrations are induced by the same agents as mutations: chemicals, radiation, viruses, common environmental agents and metabolites. Reported incidence of cells with aberrant chromosomes in blood cultures of middle-aged individuals varies in the range of 0.08 to 2.5 percent (Fischer et al., 1966; Sandberg et al., 1967; Jarvik and Kato, 1970; Kadotani et al., 1971; Kuznetsova, 1972; Sevankaev et al., 1974; Kadotani et al., 1978; Lezhava and Khmaladze, 1978; Tone et al., 1988; Bender et al., 1989; Forne, 1992; Vorobsova et al., 2001; Pilins'ka and Dybs'kyi, 2004).

Age-related rates of chromosome aberrations are dependent on a cluster of interactive factors, such as individual features, sex, cellular metabolic rate, internal milieu, DNA repair rates. Structural chromosome mutations may result from one or several breaks or linkages (deletions, duplications, inversions, translocations); they may or may not be inherited as balanced conditions (Prokofyeva-Belgovskaya, 1986). Bender et al. (1974) and Durante et al. (1996) proposed a general model for chromosome aberrations. Which suggests that (1) aberrations are induced by polynucleotide chain breaks and recombinations between disintegrated sites. Most of aberration types originate with double DNA breaks. Double breaks that occur prior to chromosome replication replicate by themselves and are present in the nearest metaphase as chromosome aberrations. Double breaks arising after partial DNA replication present as chromatid aberrations. Single polynucleotide chain breaks manifest themselves as nearest-metaphase achromatin damages or gaps; (2) the double breaks in polynucleotide chains may be directly produced by a damaging agent, but most of mutagens act either during enzymatic repair or subsequent DNA replication. DNA lesions like pyrimidine cyclobutane dimers, base abnormalities or chemical mutagen-induced cross-linkages between DNA chains are transformed into single breaks by an excision repair mechanism. Cells can repair most of such damages, but some become double breaks. Polynucleotide chain sites with these lesions cannot serve as a template during DNA replication, and gaps occur in a newly-produced DNA chain. Recombination or postreplication repair by physical exchange of terminal DNA strands is associated with damage extension to both daughter double helices.

All chromosome aberrations observable at the metaphase are interpreted as the result of the double breaks and recombinations between torn DNA strands. Every new DNA duplication cycle puts existing aberrations into a higher rank: unless metaphase achromatin aberrations are repaired, they become chromatid deletions in the second division, and give rise to chromosome deletions in the third division.

This model makes it feasible that cell exposure to chemical mutagens and ionizing radiation may produce various chromosome aberrations at different steps of the cell cycle.

Some investigators assert that DNA synthesis harbors qualitative and quantitative aberration outputs. If a chromosome is one-stranded at the time of damage (anaphase, telophase, G0 and G1 period, interphase), a rearrangement which happens to occur is defined as chromosomal: reduplication makes it such. If a rearrangement occurs after reduplication (in the G2 period) and involves one of two chromatids, it is a chromatid aberration. A postreduplication rearrangement affecting both chromatids is an isochromatid aberration (Buckton and Evans, 1975).

In a cytometric study of Staino-Coico et al. (1983), a greater percentage of thymidine-labeled phytohemagglutinin-stimulated lymphocytes from subjects aged over 65 years were found to be in the mitotic G2 and M phases, as compared to those from 20- to 30-year-old donors.

Our studies have shown that the higher incidence of aberrant karyotypes at 80 years and older is determined by chromatid and chromosome aberrations (single and paired fragments, ring-shaped and dicentric chromosomes, translocations). However, chromatid aberrations prevail over chromosomal ones. This observation agrees with evidence of Mattevi and Salzano (1975a); it might suggest that only chromatid aberrations occur in persons aged over 80.

However, there is alternative evidence that the aberration output is not actually related to a time of mutagen action (DNA synthesis or other stages) (Fesenko and Morozov, 1976). Therefore, the common view that chromosomal or chromatid versions of aberration are dependent on chromosome duplication remains questionable. Whether a chromosome or chromatid aberration will ensue depends not only on a cycle stage at which a mutagen acts but also on the magnitude of cell damage which may result in aberrations of different types (Fesenko and Morozov, 1976).

The prevalence of cells with chromatid aberrations in very old age may be related to cumulative effects of radiation and chemical mutagens that are known to induce mainly chromatid aberrations (Demoise and Conard, 1972; Sevankaev et al., 1974). The production of chromosome aberrations is a complex cascade. Some arise from early lymphopoietic divisions (Bochkov et al., 1968; Dubinin and Tarasov, 1979). Several studies showed no relationship between spontaneous aberration rates and age (Sandberg et al., 1967; Goodman et al., 1969; Bochkov et al., 1972; Sevankaev et al., 1974). A possible reason for this was a broad range of variability of spontaneous aberration rates; reevaluation of lymphocyte cultures (Bochkov et al., 1968) and follow-up controlled studies (Bloom et al., 1973) actually confirmed it.

Further, group-specific distribution of metaphase chromosome breaks showed a nonrandom pattern in blood cultures of individuals older than 80. The incidence of B and C group chromosome damage in females and A1, A2, B group and Y chromosome damage in males was above a predictable range. These patterns are consistent with reported evidence on selective chromosome aberrations associated with human aging. Thus, damages of B group chromosomes were correlated with aging (Kuznetsova, 1972) and those of A1 and C group chromosomes with mutagen exposure (Bochkov and Kuleshov, 1971; Kuleshov, 1972a, b).

Our cytogenetic studies suggest that the incidence of aberrant cells and aberrations at 80 years and older is significantly higher than at ages between 20 and 48 years.

Moreover, summing up all the results about chromosome aberrations induced by heavy metal salts, we may come to the conclusion that the effect of two metals (copper and cadmium) testing in model systems in vivo and in vitro aging is not always the same. Differences existing on the actions of these metals ions can be explained if we take into consideration the published data about the influence of the given metals on the modification of chromosome structure.

A new method detects DNA-protein cross-links induced by various agents, such as nickel, chromium and cis-platinum. It has demonstrated an increase in DNA-protein cross-links in cultured cells from humans treated with chromium and nickel. The presentage of cross-link was significantly higher among welders (1. 85 ± 1.14) than among controls (1.17± 0.46 - $P < 0.05$) (Costa, 1993).

The existing data indicate that copper ions cause conformational changes (condensation) of chromosomes when acting on the ovarian cells of Chinese hamsters. At the same time, the availability of chromosomes to the histone H2B antibodies decreased (Turner et al., 1987). As is known, the condensed chromosome regions are not available for reparation enzymes and, accordingly, on acting by factors causing chromosome condensation intensity of repair processes significantly falls (Yeilding, 1974). In its turn, this may become the reason of the rising of the mutation rate (Theophanides and Anastassopoulou, 2002; Fainaru et al., 2003).

Our data are fully placed in the represented scheme and confirm the correctness of the given assumption (Lezhava, 1996; 2001a).

However, data concerning CdCl2 as a mutagen indicate no clastogenic effect of this element (Ercal et al., 2001; Platz et al., 2002). Cadmium is an extremely toxic element (Jokhadze and Lezhava, 1994). Even following prolonged treatments, despite its cytotoxic action, there was no clastogenic effect on mammalian systems in vitro. In female mice, subcutaneous injections induced numerical but not structural aberrations in metaphase II oocytes (Watanabe et al., 1979). Acute exposure to CdCl2 for 6h led to breaks, deletion and decondensation in rodent bone marrow chromosomes in a dose-dependent and exposure dependent manner (Muramatsu et al., 1980). As ions of this element cause chromosome decondensation, their action in this sense is contrasted against the copper ions action. Effectiveness of reparation processes in this case must not be decreased; accordingly, the mutation rate must not be increased, and it was fixed in our experiment. The chromosome mutation level when acting by cadmium ions (4.0 ± 0.98%) did not differ from the control level in "old" systems (3.94 ± 1.06% in vivo and 5.25 ± 1.11% in vitro) (Jokhadze and Lezhava,1994). Therefore, the obtained data support the view that the process of heterochromatinization is primary to mutation processes.

CONCLUSION

The evaluation of spontaneous rates of structural and numerical chromosome abnormalities in the age range of 20 to 114 years has shown that very old age (80 to 114 years) is associated with an increase in the population of cells with aneuploid and aberrant chromosome sets and with a higher incidence of chromosome aberrations. The probability of natural aneuploidy is appreciably higher in females aged 80 to 108 years. Males of 80 to 114 years of age are more prone to artifactual aneuploidy. The incidence of natural aneuploidy in males remains an issue. Significant disorders in groups A, B and C chromosome counts are seen in very old age.

Chromosome damage increases in old age and its distribution makes up a nonrandom pattern. The incidence of B and C group chromosome damage was higher than predicted estimates in females of 80 years and older, while A1, A2, B group and Y chromosome damage was excessive in males. The chromatid type of aberrations prevails in both age groups.

An increasing mutation incidence induced by copper sulfate was observed in vivo and in vitro aging versus control. Cadmium chloride did not have the same effect. The difference between the effects of testing metal salt actions is suggested to be caused by their various influences on chromatin modification.

Chapter III

Repair

3.1. Repair of DNA

Repair systems that maintain the stability of hereditary information provide a major mechanism for reversal of DNA molecule damage by ultraviolet and ionizing radiation, chemical agents, viruses, and excessive temperatures. Of a variety of investigated mechanisms for DNA repair (Zhestyanikov, 1979; Zasukhina and Sinelshchikova, 1979; Small and Aronson, 1988; Pincheira et al., 1993; Weirich-Schwaiger et al., 1994; Kyng and Bohr, 2005), most detail has been elucidated for dark repair which falls into two strategies; (1) excision resynthesis, a mechanism by which DNA molecules are cut out and substituted; and (2) postreplication recovery, most of which occurs as a mechanism for recombination following the first replication cycle of damaged DNA.

Excision resynthesis, which was originally described in bacteria, has several steps. First, the excision of pyrimidine dimers from the one-helical DNA strand by cutting a sugar-phosphate bond which is a product of endonuclease activity (Setlow and Carrier, 1964). Second, partial DNA degradation which widens a gap produced by activation of exonuclease enzymes. Third, recovery of the gap with the use of complementary chain hereditary information as a template. Fourth, polynucleotide ligase-mediated splicing of the free end of the resynthesized region to a DNA molecule.

A depictor of excision repair rates is unscheduled DNA synthesis, which is a relatively low non-semiconserved synthesis induced by irradiation in non-S phase cells, regions of resynthesized material are built into an old DNA chain. Unscheduled DNA synthesis in an intact cell seems to reflect repair processes that are designed to eliminate spontaneous cytogenetic abnormalities (Zhestyanikov, 1979).

Comparison of unscheduled DNA synthesis rates in skin fibroblasts from seven mammalian species including the human has demonstrated a correlation between the rates of excision repair and the species' lifetimes (Hart and Setlow, 1976). The incorporation of labeled thymidine by cells in an extrasynthetic period was higher in mammals with longer lifetimes. The relationship between unscheduled DNA synthesis rates and species-specific lifetime gives weight to the hypothesis that the lifetime is dependent on the functional efficiency of genes for DNA repair (de Boer et al., 2002). The frequency of radioactive label

[3H]-thymidine incorporation in human lymphocyte chromosomes in the course of repair (extra) synthesis of DNA was assessed. Differential repair activity of human chromosomes was found (Stukalov, 1995).

To understand how repair intensity relates to aging, it should be determined when in the lifetime repair efficiency is impaired. Long-term cultured normal cell strains degenerate and die after a limited number of passages. In a study of Hayflick (1965), two cultures of human male diploid WI-1 cell strains in the 49th passage were mixed with a suspension of actively dividing young female WI-25 cells in the 13th passage; at 17 passages the co-culture contained only juvenile female cells that had experienced as many divisions as young control cells. In addition, numbers of divisions in a species (114-130 in tortoises, 40-60 in humans, 15-35 in chicks and 14-28 in mice) were correlated with life durations (Hayflick, 1973; Goldstein, 1974). The different caffeine effects and G_2 values between lymphocytes from old and young individuals are most likely due to a higher number of DNA lesions reaching the G_2 phase and/or a decrease of the G_2 repair capability of lymphocytes from older individuals (Pincheira et al., 1993).

It was reported by Hayflick and Moorhead (1961) that no chromosome changes were seen in human embryonic lung fibroblasts until the 40th generation inclusively, but cultures with more than 41 generations displayed increased percentages of aneuploid and aberrant cells (Saksela and Moorhead, 1963). Unscheduled DNA synthesis rates proved lowest in late (18th to 60th) passages of "young" WI-38 strain cultures treated with ultraviolet (UV) light, N-acetonal, AAF or gamma rays, and many cells showed no unscheduled synthesis (Hart and Setlow, 1976). These findings were interpreted as a loss of selected coordinatory steps in cell repair with increasing age.

The UV-induced unscheduled DNA synthesis in aging cultured human fibroblasts was first noted by Goldstein (1971). Exposure of the terminal fibroblast cultures to different dosages of UV irradiation depressed the synthesis. A similar experience with ionizing and UV irradiation was reported by many others (Little, 1976; Roth et al., 1989).

Little et al. (1974) employed equilibrium gradient centrifugation to assess the repair of X-ray induced one-strand DNA breaks in human skin fibroblast cultures. The repair was significantly reduced in aging late-passage cells (before death of one or two duplications in the cell population).

Wheeler and Lett (1974) identified more spontaneous breaks in DNA of 'old' cells from the granular cerebellum of 13-year-old dogs than in DNA of 'young' cells from 7-week-old dogs. Zhestyanikov and Vilenchik (1978) have proposed two explanations for the higher incidence of one-strand DNA breaks in aging cells. The breaks may accumulate because of age-related repression or deficiency of repair enzymes that are required to mend the DNA breaks (DNA polymerase I and III and DNA ligase). Probably, it is not incidental that the one-strand break accumulation is the first DNA lesion recognizable in aging cells since this lesion is commonly seen in normal cellular activities (replication, transcription and recombination).

Mattern and Cerutti (1975) have examined the excision of 5, 6-dioxydihydrotinin, a mechanism which cultured fibroblasts use to repair gamma irradiation-induced damage, and found out that old fibroblasts were less efficient in DNA splicing. It has been demonstrated that aged *in vitro* hepatocytes have lower repairability regardless of their sensitivities to employed damaging agents—biomycin or UV light (Kennah et al., 1985). The ability of cells to repair DNA strand lesions is lost with both premature (progeria) and normal aging (Wheeler and Lett, 1974; Weirich-Schwaiger et al., 1994). The lower rates of DNA repair synthesis have been documented in lymphocytes from humans over 60 years of age exposed to UV irradiation (15 J/mm^2) (Lambert et al., 1977).

However, there is a line of evidence suggesting that aging does not affect cell repair. In a study of Painter et al. (1973), the repair rates of UV-irradiated cells remained similar at the 20th and 41st passages. Human diploid WI-38 fibroblasts normally repaired UV-damaged DNA, and the repair rates were comparable in young (passages 15 to 29) and old (passages 46 to 55) cells (Clarkson and Painter, 1974).

3.1.1. Repair in Human Congenital Diseases

Several types of mutations in humans may present as severe congenital diseases, in which a genetic defect occurs as abnormal DNA repair or excessive spontaneous chromosome aberrations. Michelson (1986) has suggested that the best known of these diseases be categorized into three groups.

Group I comprises diseases that are collectively referred to as Xeroderma Pigmentosum (XP). XP is underlied by an autosomal recessive mutation and manifests in early childhood as capillary dilation, keratosis, abnormal speckle formation and excessive cutaneous sensitivity to sunlight, which typically leads to a malignancy. Serious neurological disorders are seen. The patients rarely survive to the age of 30. Ectodermal and mesodermal malignant tumors are most common and in patients with XP variant forms I and II chromosome aberrations occur at a higher rate (Michelson, 1986; Tang et. al., 2000).

Group II are diseases with a collective name fragile chromosome syndrome. It includes Fanconi's anemia, Bloom's syndrome and Ataxia-telangiectasia (Louis-Bar syndrome) (Bohr et al., 1989).

Fanconi's anemia (FA) is a hereditary disease. FA is a autosomal recessive syndrome associated with chromosomal instability, hypersensitivity to DNA cross-linking agents, and predisposition to malignancy. The gene for FA complementation group G (FANCG) was the third FA gene to be cloned, and was found to be identical with human XRCC9, which maps to 9p13. The cDNA is predicted to encode a polypeptide of 622 amino acids, with no sequence similarities to any other known protein or motifs that could point to a molecular function for FANCG/XRCC9. The Portuguese-Brazilian, French-Acadian, and Korean/Japanese mutations were likely to have been present in a founding member of each of these populations (Aurbach et al., 2003).

In order to characterize the molecular defects underlying FA in Tunisia, 39 families were genotyped with microsatellite markers linked to a known FA gene. Haplotype analysis and homozygosis mapping assigned 43 patients belonging to 34 families to the FAA group, whereas one family was probably not linked to the FANCA gene or to any known FA genes. For the patients belonging to the FAA group, screening for mutations revealed four novel mutations: two small homozygous deletions 1693delT and 1751-1754del, which occurred in exon 17 and exon 19, respectively, and two transitions, viz., 513G—>A in exon 5 and A—

>G at position 166 (IVS24+166A—>G) of intron 24. Two new polymorphisms were also identified in intron 24 (IVS24-5G/A and IVS24-6C/G) (Bouchlaka et al., 2003).

Bloom's syndrome is a very rare autosomal recessive congenital disease. It occurs as immune dysfunction, hypersensitivity to solar irradiation and susceptibility to malignant neoplasms. It is commonly associated with a higher incidence of spontaneous and induced chromosome aberrations and lower cell survival (Ellis et al., 1995).

Ataxia-telangiectasia (AT) is an autosomal recessive disease with neurovascular involvement. Major presentations of AT are loss of motor coordination, immune impairment and susceptibility to lymphoid and other malignancies .The patients usually die at ages of 12 to 15. The patients' cells show increased percentages of chromosome aberrations and excessive sensitivity to irradiation (Martin et al., 2003). Significant impairment has been found in rejoining of gamma ray-induced one-strand DNA lesions in fibroblasts from AT patients (Zdzienicka et al., 1994).

Group III includes diseases that are defined as premature aging; most prominent of these is progeria. This hereditary disease causes repair disorders. There are child's progerias, or the Hutchinson-Gilford syndrome (HGP), and adult progeria, or Werner's syndrome. The Hutchinson-Gilford syndrome is a rare genetic disease. Already in early postnatal months affected infants have a senile appearance and show loss of subcutaneous fat and hair, and prominent atherosclerosis involving the brain and heart. Adolescent patients look very senile and die before the age of 20 (Martin and Oshima, 2000).

The appearance of Proline/arginine-rich end leucine-rich repeat protein (PRELP) sometime after the third month of the birth coincides with the appearance of HGP symptoms. HGP has been diagnosed in twins with a chromosomal inversion at, or very near, the site of the PRELP gene (Lewis, 2003).

Unscheduled DNA synthesis rates were reported to be similar in UV-irradiated fibroblasts from progeria patients and normal subjects. Subsequent studies have indicated abnormality of DNA repair synthesis in progeric cells and in neurodegenerative diseases. In the study of DNA repair was stimulated in progeric EX/17 fibroblasts as they were co-cultured with normal human diploid early- or mid-passage fibroblasts. However, no such stimulation occurred as the progeric cell lines were co-cultured with old diploid late-passage fibroblasts. Therefore, DNA repair defects and premature aging are recognizable in the progeric cell lines (Brown et al.1976; Weirich-Schwaiger et al., 1994).

Down's syndrome is another disease associated with premature aging, chromosomal abnormalities (Barenfeld, 2002) and defects of repair enzymes (Takeshita et al., 1992). Abnormality may occur as trisomies (47, XX or XY, 21+) and translocation, or it may be mosaicism, when cells with trisomy 21 coexist with cells having the normal chromosome number of 46.

Patients with Down's syndrome show signs of premature aging already at ages of 15 to 40: a dry wrinkled skin, joint stiffness, tooth loss, central nerve disorders. Trisomic cells of the patients are susceptible to malignant transformation; the incidence of leukemia in such patients is 18-20 times as high as in normal populations. Frequencies of both chromosome-type exchanges and deletions were elevated in the patients with Down syndrome by about 1.3 times in comparison with controls (Takeshita et al., 1992). Repair studies in lymphocytes from patients with Down's syndrome have revealed an impairment of unscheduled DNA synthesis after exposure to UV light (15 J/mm2) (Lambert et al., 1977; Saadat et al., 1998; Boer et. al. 2002).

In summary, the literature presents a hypothesis that in vivo and in vitro cell aging in mammal animals and humans is related to deterioration of DNA repair. Evidence for a lower efficiency of repair processes in aging and culture cells lends support to it. On the other hand, some studies have questioned the correlation between aging and repair decline.

3.2. Unscheduled DNA Synthesis

Unscheduled DNA synthesis has been investigated in UV-irradiated lymphocytes from very old subjects of both sexes; the study group involved 10 subjects aged 80 to 90 years, and the control group included 19 subjects aged 20 to 43 years (lymphocyte counts of 80- to 90-year-old individuals were within the middle-age range; Kipshidze et al., 1964). Oxyurea (10 mmole per ml) of the culture was used in all study series as an inhibitor of DNA replicative synthesis (even in the absence of stimulation about 0.1 percent of lymphocytes were reported to have the replicative synthesis; Evans and Norman, 1968). Therefore, DNA synthesis registered was unscheduled repair synthesis. The cultures were exposed to UV irradiation, at 10-15 J/mm^2, using a bactericidal lamp (λ = 254 nm; capacity, 1.98 J/mm^2/s).

Immediately after irradiation [3H]-thymidine was introduced into the medium (final concentration, 10 µCi/ml; specific radioactivity, 14 Ci/mmole). The cultures were incubated in centrifuge tubes for 2.5 hours at 37°C and then centrifuged at 800 g for 5 min at room temperature. Supernatants were removed, and 0.5 ml of Haenks' solution was added to the sediment. Cells were carefully resuspended and transferred onto filters. The filters were dried, washed thrice with 5 percent trichloracetic acid cooled to 4°C, rinsed with absolute alcohol and dried. Radioactivity was measured in 5 ml of the toluol scintillation fluid using a Nuclear Chicago Mark-2 counter (USA).

Unscheduled DNA synthesis was low in all nonirradiated cells from the study subjects (aged 80 to 90 years) and the control group (20 to 43 years). The rates of unscheduled DNA synthesis induced by UV irradiation varied within a wide range in both groups (Fig.3.2-1). The synthesis in the study and control cells was highest after exposure to 15 J/mm^2. Higher irradiation dosages occasionally employed by us (20-35 J/mm^2) proved less efficient.

Unscheduled DNA synthesis rates at old ages (Fig. 3.2-1) were considerably lower when UV dosages of 10 to 15 J/mm^2 were applied: exposure to 10 J/mm^2 produced a rate of 239 ± 152 count per minute (cpm) at old ages versus 417 ± 148 impulses per minute at middle ages. Respective values for exposure to 15 J/mm^2 were 249 ± 183 and 456 ± 165 cpm.Therefore, lymphocytes from humans older than 80 exhibits a substantial reduction in unscheduled DNA synthesis induced by UV irradiation.

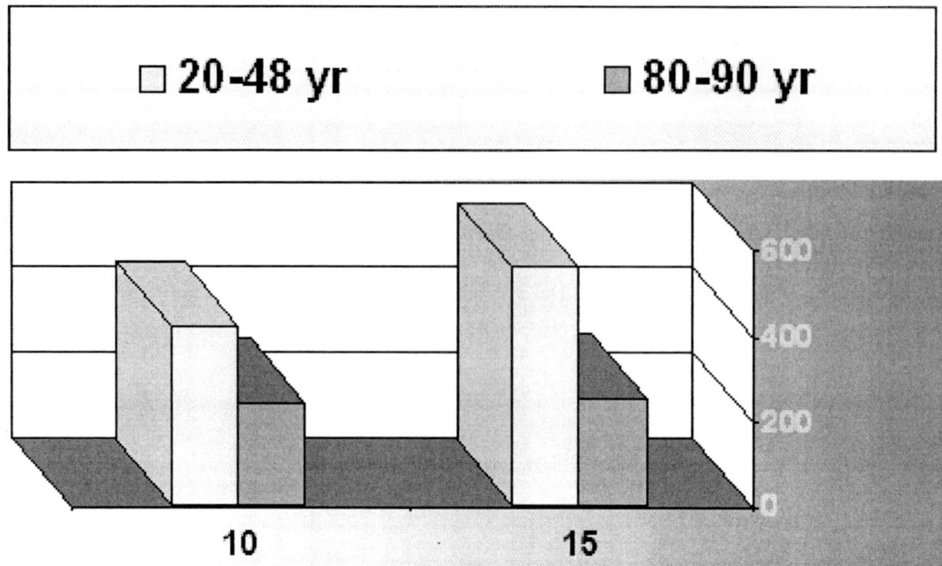

Fig. 3.2-1. Unscheduled DNA synthesis induced by ultraviolet irradiation in lymphocytes of subjects aged from 80 to 90 yr.

3.3. Sister Chromatid Exchanges

Sister chromatid exchange (SCE) is an isolocus exchange between sister chromatids with a necessary duplex rejoining. It is a DNA repair strategy which fails with cell and body aging (Schneider et. al., 1979).

Sister chromatid exchanges were postulated by McClintock in 1938, as she evaluated dicentric ring-shaped chromosomes in maize. Sister chromatid crossingover at anaphases I and II of a meiotic ring-shaped chromosome was identified in heterozygous maize by Schwartz in 1953. The presence of SCE was documented by Allen et al. (1976) in a murine mitotic metaphase I (unpaired sex chromosome).

The first detailed study of SCE was undertaken by Taylor et al. (1958). 3H-thymidine radioactivity was recovered during the first division in both metaphase I sister chromatids, while the label was recognized in only one chromatid during the second-division metaphase II, suggesting a semiconserved replication. Moreover, sister chromatids were found to exchange material since the label of one was recovered in the other.

The results of Taylor et al. were met with skepticism (Wolff, 1964); it was claimed that SCE were not spontaneous but induced by the exogenous radioactive label. However, the studies of Brewen and Peacock (1969) and Gibson and Prescott (1972) gave firm evidence for spontaneous SCE. Later, differential staining techniques were developed for SCE evaluation. Zakharov and Egolina (1972) first demonstrated that a thymine analog, unlabeled 5-bromodeoxyuridine (BrDU), interferes with chromatid condensation. These researchers employed BrDU in cell cultures and found a higher SCE incidence during two-cycle chromosome reproduction.

Refinements in Giemsa and fluorochrome staining techniques for identification of brominated DNA in chromatids gave a new incentive for SCE studies (Perry and Wolff, 1974). Current unravelling of SCE mechanisms is a major problem in molecular biology. Success with SCE identification has been achieved with the use of the 33258-Hoechst-Giemsa fluorescent stain (Perry and Wolff, 1974; Fischer and Kim, 1975) or techniques using no fluorochromes (Antoshchina and Poryadkova, 1978).

The frequency of SCE is known to increase with greater BrDU concentrations (Kato, 1974a, b). A SCE number per metaphase at a middle age varies from 4.1 to 41.6 (Beek and Obe, 1975; Lambert et al., 1976). An average is 8-10 SCE per metaphase (Sperling et al., 1975; Tice et al., 1975; Bairamyan, 1978). No intersubject and intersex variability has been reported in the incidence of SCE (Crossen et al., 1977; Bairamyan, 1978).

Studies in early-, mid- and late-passage cells, lymphocytes and fibroblasts from human donors and animals of different ages (in vivo and in vitro) have found either no change of lymphocyte SCE incidence with age (Galloway and Evans, 1975; Morgan and Crossen, 1977; Schneider et al., 1979; Vormittag, 1983; Bender at al., 1989; Bukvic et al.,2001) or an increase with progressive age (Wen Wu-Nan et al., 1983; Bukvic et al.,2001). However, the SCE incidence in human fibroblasts and animal peripheral lymphocyte cultures has been reported to decrease with age and with the greater number of cell passages (Schneider et al., 1979; Ardito et al., 1983). In general, chromosome-specific SCE distribution is proportional to chromosome lengths and DNA contents (Galloway and Evans, 1975; Crossen et al., 1977; Bairamyan, 1978). However, E, F and G group chromosomes showed SCE numbers that were less than expected (Crossen et al., 1977; Bairamyan, 1978).

Research addressed SCE locations in hetero- and euchromatin. Sperling et al. (1975) showed that SCE rates are lower in centromeric and pericentromeric regions, where constitutive heterochromatin resides. Crossen (1983) has detected less SCE in the heterochromatin region 9gh+ of the human chromosome 9, as compared to euchromatin. Evidence from other studies has confirmed the lower SCE rates in constitutive heterochromatin regions (Galloway and Evans, 1975; Hsu and Pathak, 1976; Tada-Aki Hori, 1983). Latt (1974) found SCE to locate mostly in low-fluorescent chromosome regions and less commonly near brightly fluorescent ones. SCE were also identified at the heterochromatin-euchromatin junction in Indian Munjac chromosomes (Carrano and Wolff, 1975) and in euchromatin regions of Chinese hamster and *Microtus montanus* chromosomes (Hsu and Pathak, 1976). However, there is controversial evidence suggesting higher SCE rates in heterochromatin regions (Tice et al., 1975; Holmquist and Comings, 1975; Lin and Alfi, 1976). The frequency of SCE is particularly high within the subtelomeric regions of chromosomes (Conforth and Fleerle, 2001, Bailey et al., 2004).

Table 3.3-1 summarizes the data of the SCE study in lymphocytes cultured from subjects of 80 to 88 years of age and controls (Lezhava, 1987; 2001a,b).

A total of 5,510 exchanges were detected in 677 metaphases (one-metaphase mean, 8.28; range, 2 to 23). One-metaphase mean ± SD varied from 6.53 ± 1.05 to 15.80 ± 2.95. Of 29 subjects, 17 had SCE counts that significantly deviated from mean values. However, some families showed no individual variability of SCE counts (Table 3.3 -2).

Individual variability of SCE rates has been explained by technical factors and study designs (Bairamyan, 1978). Of special importance is a concentration and duration of the presence of BrDU in a growing culture. In our study, the same BrDU concentration (7.7 μg/ml) was used in both age groups, and a bromide analog was introduced for 96 hours as the culture was established.

The SCE data for individuals from the two age groups were arranged into the following set:

$$
\begin{array}{ll}
X_{1,1}, X_{1,2}, \dots, X_{1,n_1}, & Y_{1,1}, Y_{1,2}, \dots, Y_{1,m_1}, \\
X_{2,1}, X_{2,2}, \dots, X_{2,n_2}, & Y_{2,1}, Y_{2,2}, \dots, Y_{2,m_2}, \\
\dots\dots & \\
\dots\dots & \\
X_{N,1}, X_{N,2}, \dots, X_{N,n_N}, & Y_{M,1}, Y_{M,2}, \dots, Y_{M,m_M}
\end{array}
$$

where$X_{1,1}, X_{1,2}, \dots, X_{1,n_1}$ is an observed SCE value in $n1$ cells of the first individual from the middle-aged group, $X_{2,1}, X_{2,2}, \dots, X_{2,n_2}$, is a value of the second individual (n_2 is a number of assayed cells) and so on; N stands for a total number of individuals. Likewise, Y denotes SCE counts of old subjects (M is a total number of the individuals).

Our comparison of the two age groups was based on the assumption that a mean SCE count varies with subjects within the age groups (Fig. 3.3.-1); however, SCE means are also distributed intersubjects in each group, and so it is the average SCE values for the groups that should be compared.

Such a protocol provides a test of difference

$$
t = \frac{\overline{X} - \overline{Y}}{\sqrt{S_{\overline{X}}^2 + S_{\overline{Y}}^2}}, \quad \overline{X} = \frac{1}{n}\sum_{i=1}^{N}\sum_{j=1}^{n_i} X_{i,j}, \quad \overline{Y} = \frac{1}{m}\sum_{i=1}^{M}\sum_{j=1}^{m_i} Y_{i,j};
$$

$$
n = \sum_{i=1}^{N} n_i, \quad m = \sum_{i=1}^{M} m_i,
$$

$$
\overline{S}_{\overline{X}}^2 = \frac{1}{n} S_{1,1}^2 + \frac{\sum_{i=1}^{N} n_i(n_i - 1)}{n^2} S_{1,2}^2, \quad \overline{S}_{\overline{Y}}^2 = \frac{1}{m} S_{2,1}^2 + \frac{\sum_{i=1}^{M} m_i(m_i - 1)}{m^2} S_{2,2}^2,
$$

$$
S_{1,1}^2 = \frac{1}{n}\sum_{i=1}^{N}\sum_{j=1}^{n_i}\left(X_{i,j} - \overline{X}\right)^2, \quad S_{2,1}^2 = \frac{1}{m}\sum_{i=1}^{M}\sum_{j=1}^{m_i}\left(Y_{i,j} - \overline{Y}\right)^2,
$$

where N, M are numbers of individuals in the age groups; ni ($i = 1, \dots, N$) is a number of assayed metaphases from middle-aged individuals; mi ($i = 1, \dots, M$) is a number of assayed metaphases from old individuals; Xi, j is a SCE count at the jth metaphase of the ith middle-aged individual ($1iM$, $1\ jni$), Yi,j is a SCE count at the jth metaphase of the ith old individual; is a SCE mean for the ith middle-aged individual; is a SCE mean for the ith old individual; is a SCE mean for a middle age; is a SCE mean for an old age; is dispersion of Xi, j SCE values in the middle-aged group; is dispersion of Yi,j SCE values in the old-aged group; is interindividual dispersion of mean *Xi* values within middle age; is dispersion of mean values within old age.

Table 3.3-1. SCE counts in peripheral lymphocytes from individuals aged 18 to 59 years and 80 years or older

Age (yr)	No. of metaphases	SCE counts	SCE count per metaphase (mean ±SD)	SCE counts in different chromosomes or chromosome groups								
				A_1	A_2	A_3	B	C	D	E	F	G
18	41	305	7.44 ± 0.97	35	40	20	41	119	32	14	3	1
20	20	156	7.80 ± 1.82	27	13	7	14	71	18	4	2	—
23	25	238	9.52 ± 1.22	23	23	20	30	119	20	3	—	—
25	50	344	6.88 ± 0.55	35	27	30	55	149	0	10	4	4
26	15	98	6.53 ± 1.05	15	10	6	13	38	9	6	1	—
31	15	118	7.87 ± 2.09	22	9	6	13	51	10	7	—	—
40	25	193	7.72 ± 1.38	20	11	10	17	105	17	7	5	1
42	50	434	8.68 ± 0.92	50	51	29	59	191	38	13	1	2
44	15	131	8.73 ± 1.78	10	15	13	17	51	11	9	1	4
50	30	239	7.97 ± 1.19	25	19	9	30	112	20	9	2	3
51	22	214	9.73 ± 0.85	22	30	10	27	88	22	9	4	2
52	10	158	15.80 ±2.95	13	16	15	29	63	14	6	1	1
53	30	316	10.53 ± 1.43	40	28	26	41	136	30	13	1	1
55	21	157	7.48 ± 1.15	24	14	9	27	61	15	5	1	1
59	26	189	7.27 ± 1.09	36	17	12	22	67	22	11	2	—
59	35	320	9.14 ± 1.11	39	30	29	46	128	32	15	—	1
Total	430	3610	8.40 ± 0.33	436	353	251	481	1549	350	141	28	21
80	30	250	8.33 ± 1.06	20	28	17	38	102	23	15	7	—
80	15	100	6.67 ± 2.10	10	11	4	18	38	14	3	2	—
80	15	110	7.33 ± 0.93	9	12	7	29	38	4	11	—	—
80	10	111	11.10 ± 2.49	7	9	9	8	61	13	2	1	1
80	20	128	6.40 ± 0.97	8	12	12	9	66	14	6	—	1
80	21	177	8.43 ± 1.04	18	23	15	24	67	20	9	—	1
80	19	144	7.57 ± 1.27	12	17	14	27	51	15	3	3	2
81	23	156	6.78 ± 0.84	9	17	8	41	59	10	11	1	—
83	27	216	8.00 ± 1.10	20	31	8	27	100	14	12	2	2
83	10	71	7.10 ± 1.07	8	8	6	10	24	9	4	2	—
87	11	84	7.64 ± 2.02	12	10	4	16	27	9	6	—	—
87	26	208	8.00 ± 0.68	15	28	14	27	79	25	13	3	4
88	20	145	7.25 ± 0.45	12	12	11	30	57	17	5	1	—
Total	247	1900	7.69 ± 0.34	160	218	129	304	769	187	100	22	11

Table 3.3-2. Significance of SCE variability in family members

Families	Age (yr)	SCE counts per metaphase (mean±SD)	P
Mother	80	7.33 ± 0.93	> 0.05
Father	88	7.64 ± 2.02	> 0.05
Son	59	7.27 ± 1.09	
Mother	80	11.10 ± 2.49	> 0.05
Father	87	8.00 ± 0.68	< 0.001
Son	52	15.80 ± 2.95	
Mother	80	8.33 ± 1.06	> 0.05
Son	44	8.73 ± 1.78	
Father	87	7.64 ± 2.02	> 0.05
Son	59	9.14 ± 1.11	
Father	83	8.00 ± 1.10 <	< 0.01
Son	51	9.73 ± 0.85	

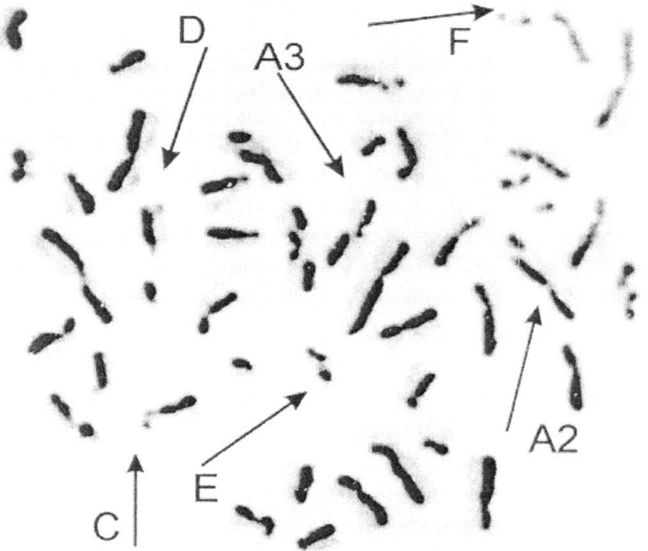

Fig. 3.3-1.Differential staining of sister chromatids after culture treatment with BrDU. Arrows indicate SCE in chromosomes 2,3, C,D,E,F (lymphocyte culture).

Application of Student's test without controlling for intra age dispersion

$$t = \frac{\overline{X} - \overline{Y}}{\sqrt{\frac{S_{1,1}^2}{n} + \frac{S_{2,1}^2}{m}}},$$

yields $t = \frac{8.40 - 7.69}{\sqrt{0.028 + 0.03}} = \frac{0.71}{\sqrt{0.058}} = \frac{0.71}{0.24} = 2.95.$

Therefore, the difference between the age groups acquires a higher significance. However, intra age variability should be controlled for, and a larger expression needs be used as dispersion for:

$$S_X^2 = \frac{S_{1,1}^2}{n} + \frac{\sum_{i=1}^{N} n_i(n_i - 1)}{n^2} S_{1.2}^2.$$

Computation of the second entry of the expression shows that it makes a substantial portion of dispersion X. = 0.028 + 0.115 = 0.143 for the middle-aged group, and = 0.03 + 0.107 = 0.137 for the old-aged group; the t criterion takes on the value of 1.4, and this testifies the intergroup difference with significance of 18 percent (Table 3.3-1).

To obtain more detail on the tendency of SCE rates to decrease with aging, we analyzed SCE data for individual chromosomes. Again, intra age variability accounts for most of dispersion of SCE means. A notable difference between control and study SCE means for the $A1$ and C chromosomes presented itself as respective values:

$$t_A = \frac{1.05 - 0.79}{\sqrt{0.0040 + 0.0047}} = \frac{0.26}{0.09} \approx 2.9$$

$$t_C = \frac{3.7 - 3.11}{\sqrt{0.020 + 0.021}} = \frac{0.6}{0.22} \approx 3$$

Here we excluded findings of subject 12 of the control group and subject 4 of the senile group since their C-chromosome SCE counts strongly deviated from group means. But even when these were controlled for, we found

$$t_C = \frac{0.52}{\sqrt{0.038 + 0.060}} = \frac{0.52}{0.31} \approx 1.7$$

Sister chromatid exchanges were identified in almost all chromosomes—from A to G groups (Fig.3.3 - 1). The analysis of chromosome distribution of SCE showed inadequate SCE in chromosomes E, F and G in both age groups. SCE counts were significantly above theoretical estimates in chromosomes A1 and C in middle age (Fig.3.3-2) and in chromosomes A2 and B in old age.

The evaluation of SCE distribution over chromosome A1, A2 and A3 lengths showed that lymphocyte SCE counts in the study and control groups were highest on chromosome medial arms and abruptly decreased in the direction of the telomere and centromere, implying that SCE rates are lower in pericentromeric and telomeric heterochromatin regions.

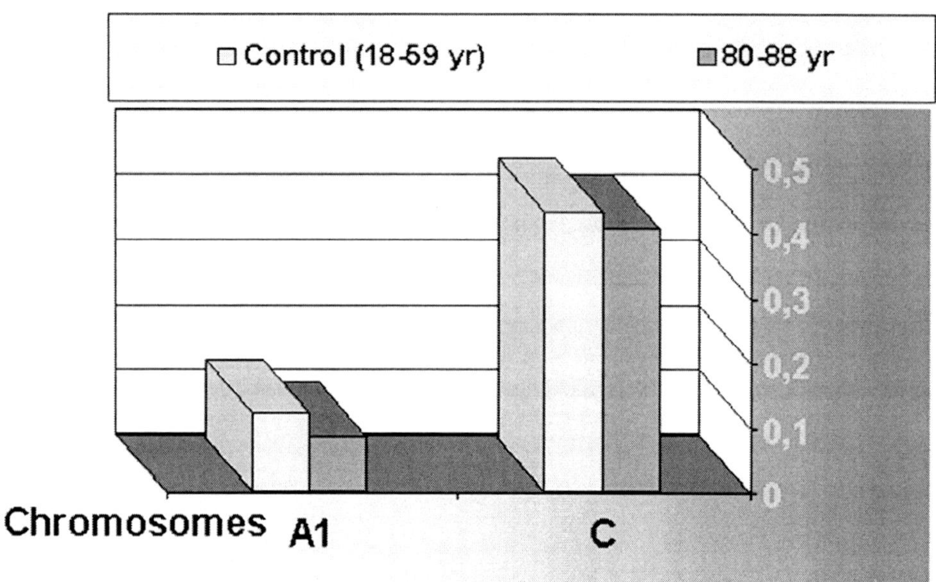

Fig.3.3-2. SCE distribution in A1 and C chromosomes of peripheral lymphocytes cultured from individuals aged 18 to 59 years and 80 years and older

DISCUSSION

Unscheduled DNA synthesis. During recent years, more consensus has been generated about the role of repair enzymes in mechanisms of aging. The aging-repair relationship emerges from evidence that (1) there are diseases that affect repair, with clinical premature aging; and (2) repair activity is impaired in cells of very old humans and old animals, and in increasingly passaged (aging) diploid cells (Zasukhina and Sinelshchikova, 1979; Vijg and Knook, 1987). Alternative evidence has challenged the correlation between age and impairment of cell repair, in particular after UV irradiation (Little, 1976; Evans and Bohr, 1994).

Nucleosomal DNA is repaired at a rate of about 10% of that for naked °*ivo* studies of mammalian cells have revealed that the organization of DNA within chromatin has a strong negative effect on its repairability by the nucleotide excision repair system. Similarly, *in vivo* studies have shown that transcribed DNA is repaired preferentially, and since transcription is invariably associated with significant chromatin remodeling, it has been inferred that the various activators, coactivators, and remodeling and accessibility factors which play essential roles in transcription may play equally prominent roles in excision repair (Hara et al., 2000; Kyng and Bohr, 2005). Studies in recent years have established that chromatin plays a central role in modulation of DNA-dependent processes. The studies on nucleotide excision repair (NER) of UV lesions provide clear examples for mutual interactions and functional links between chromatin structure, transcription and DNA repair processes. Repair of UV lesions by NER is just one example of cellular reactions to environmental mutagens (Thoma, 1999).

Unscheduled DNA synthesis is detected in human lymphocytes at UV exposure of 0.2 J/mm^2 and reaches a maximum at 4 J/mm^2, to remain at an established level until exposure to 560 J/mm^2 (Clarkson and Evans, 1972). It was reported that UV-induced repair in HeLa cells increased as UV dosages were built up to 40 J/mm^2 (Cleaver, 1974). Our experience is similar to the formerly mentioned findings, although DNA repair was occasionally seen to decrease already at UV exposure of 15 J/mm^2.

A study of unscheduled DNA synthesis revealed that 90 percent of UV- and gamma-induced lesions are repaired within 1 hour and 10 percent within 5 hours (Spiegler and Norman, 1970). Our UV-irradiated cultures were incubated at 37 °C for 2.5 hours.

According to Wheeler and Lett (1974), the inability of cells to repair structural DNA damage qualifies an organism as progeroid; other causes are massive irradiation of mammalian cells and natural cell aging. Unscheduled DNA synthesis was found to be significantly reduced in Down's syndrome, actinic keratosis and patients older than 60, as was indicated by a lower DNA thymidine incorporation in the presence of oxyurea and lower repair rates at UV exposure (15 J/mm^2) (Lambert and Ringborg, 1976; Lambert et al., 1977; Saadat et al.,1998). The functional change of cells with age was thought to occur because of somatic mutation accumulation. This agrees with our observations: unscheduled DNA synthesis at UV exposures of 10 and 15 J/mm^2 was very low in lymphocytes from individuals aged 80 or older.

It has been reported by Nette et al. (1984) that increasing age does not affect the numbers of cells maintaining repair synthesis, but synthesis intensity is lowered by an average of 50 percent. The declining ability of cells to reverse DNA lesions and a secondary increase in chromosome aberration rates in aging cells (Lezhava and Khmaladze, 1978; Hardner et al., 1982; Staino-Coico et al., 1983; Lezhava, 1991; Ramsey, 1995) result from heterochromatinization of genomic sites at older ages. Indeed, our studies have identified progressive chromosome heterochromatinization as a concomitant of aging (Lezhava et al., 1979a; Lezhava, 1977, 1980, 1996; 2001a). It is consistent with the Yielding hypothesis (1974) that repair is active only at transcriptional DNA sites; the heterochromatin portion of the genome does not repair damage and accumulates it. Autoradiographic and electron microscopic studies of Harris et al. (1974) demonstrated that unscheduled DNA synthesis in UV-irradiated human cell nuclei is twofold higher in euchromatin regions, as compared to heterochromatin ones. This suggests that the lower DNA synthesis observed by us in lymphocytes of very old subjects was secondary to heterochromatinization.

Sister chromatid exchanges. Chromatid exchanges were discovered when distribution of [3H]-thymidine label between sister chromatids was examined in successive mitoses (Taylor, 1958). Cytological tests for interchromatid exchanges confirmed the findings of Taylor.

This research was continued by Egolina and Zakharov (1972). Their studies showed that treatment of cultured Chinese hamster cells with 5-bromodeoxyuridine produced selective staining of sister chromatids after two replication cycles, and these sites often proved SCE. These studies became a starting point for all later SCE research.

The main marker of SCE during one or several DNA replication cycles is the incorporation of 5-bromodeoxyuridine instead of normal thymine, of which 5-BrDU is an analog, at second-division metaphase. Employed haloid-substituted precursors include 5-iododeoxyuridine (Ikushima and Wolff, 1974), acridine orange (Kato, 1974a; Perry and Wolff, 1974), 4-6-diamidino-2-phenylindole (Lin and Alfi, 1976), 5-bromodeoxycytidine (Bairamyan and Zakharov, 1978); fluorochromes are used in a combination (33258-Hoechst and Giemsa stain) (Perry and Wolff, 1974).

A simplified technique for cytological identification of brominated DNA in chromosomes of cultured cells is culture exposure to a mercury vapor lamp without fluorochrome staining (Antoshchina and Poryadkova, 1978).

The occurrence of SCE largely varies with BrDU concentrations. A minimal concentration (0.1 to 1 μg/ml) producing differential chromatid fluoroscence does not alter SCE and chromatid aberration identification rates (Kato, 1974b). BrDU concentrations of 50 to 200 μg/ml cause a much higher incidence of sister chromatid exchanges or incomplete chromatid condensation (Lambert et al., 1976). The BrDU concentration used by us (7.7 μg/ml) allowed a good differentiation of chromatids from scarce BrDU-induced SCE and revealed an almost spontaneous exchange rate (Latt et al., 1975; Crossen et al., 1977; Morgan and Crossen, 1977; Zakharov, 1978). BrDU may be used for two mitotic cycles and, in lymphocyte cultures, for the entire incubation period (72 or 96 hours) (Antoshchina and Poryadkova, 1978). In the experiment, BrDU introduction for 96 hours was associated with a 40 to 50 percent incidence of metaphases showing asymmetrical sister chromatids.

The literature indicates interdonor and intraobserver variability of SCE occurrence. Moreover, mean counts of SCE vary in different studies from 4.1 to 27.3 (Beek and Obe, 1975; Solomon and Bobrow, 1975). In our study, means ± SD fluctuated from 6.53 ± 1.05 to 15.80 ± 2.95 at ages of 18 to 59 years and from 6.40 ± 0.97 to 11.10 ± 2.49 at 80 years and older. It has been stated that individual genotypes explain the interindividual difference of SCE findings (Galloway and Evans, 1975; Crossen et al., 1977). This has been proven by a survey of repeat studies that showed only a relative stability of SCE counts in a given individual. The individual SCE counts may depart from a normal distribution, suggesting a scatter of SCE rates in the population (Zakharov, 1978).

Cytogenetic studies have demonstrated SCE incidence to be distinctly related to chromosome lengths, inconsistent with a theoretical range and lower in group E, F and G chromosomes (Galloway and Evans, 1975; Morgan and Crossen, 1977; Crossen et al., 1977; Bairamyan, 1978). Our findings in middle-aged and old subjects support these observations. Inadequate SCE in the small-chromosome groups appear to be determined by high heterochromatin levels (Bairamyan, 1978). Furthermore, SCE are more likely to be overlooked in the small chromosomes since their telomeric and pericentromeric regions make big portions of their lengths (Chebotarev, 1979). Our studies showed that SCE values were excessive in A1 and C group chromosomes in middle age, A2 and B group chromosomes in old age and E, F and G group chromosomes in both age groups. Excessive SCE in large chromosomes have been reported by Alhadeff and Cohen (1976) and Crossen et al. (1977). It might be related to euchromatin content per unit of chromosome length.

A relationship of SCE incidence per cell to age is of special interest. Comparison of SCE rates in lymphocytes from 8 children under 11 years and 7 adults of 26 to 35 years revealed no effect of age on SCE outputs (Galloway and Evans, 1975). Morgan and Crossen (1977) found no variability of SCE incidence in lymphocytes from 16 children under the age of 15, 24 adults of 16 to 60 years and 10 subjects of 61 to 85 years of age. No correlation was seen between the SCE incidence and age, including an extremely old one, in 72-hour lymphocyte cultures of 100 donors and 16 centenarians (Hardner et al., 1982; Buckvic et al., 2001).

Schneider et al. (1979) were the first to focus on the age-related variability of SCE incidence. They found out that the incidence decreased in human fibroblasts as age and number of passages increased. However, lymphocytes from middle-aged and old subjects did not display an age-related increase in spontaneous or mitomycin-induced SCE. Interpretations of Schneider et al. were: (1) human peripheral lymphocytes in vivo are at the interphase which is a prereplication stage; (2) antigens like phytohemagglutinin are required for stimulation of in vitro DNA synthesis; (3) SCE-inducing effects of mutagens in the lymphocytes are much different from those in fibroblasts.

However, the study of Arce (1981) has revealed a lower SCE incidence in spontaneously cultured lymphocytes from individuals of 65 years or older. The centenarian (16 subjects) of lymphocytes SCE per cell frequencies no change ($F=2{,}45$; $p<0{,}05$) (Bukvic et al., 2001). Studies of Kram et al. (1978) and Kong (1988) have demonstrated that mitomycin C-induced in vivo SCE rates in marrow cells of old mice and rats were significantly below those in young animals. Bone marrow cells from young and old C57BL/6J mice have been evaluated for SCE induction by cyclophosphamide, mitomycin C and doxorubicin. Higher concentrations of the three mutagens produced lower SCE rates in old cell populations (Nakanishi et al., 1979; Bond and Sing, 1988).

The study in the C57BL/6J mice and Wistar rats did not identify a relation of spontaneous SCE rates to animals' age. However, mutagen-induced SCE in the bone marrow and splenic cells of old animals (24 months) occurred at significantly lower rates ($P < 0.01$) than in young cells (Kram et al., 1978; Schneider et al., 1979; Bond and Sing, 1988). A similar decrease in SCE was exhibited by old cell populations treated with cyclophosphamide and adriamycin (Schneider et al., 1979).

Our studies in lymphocytes from subjects of 80 to 88 years of age have shown a tendency of SCE counts to decrease with age. The analysis of SCE in individual chromosomes revealed a prominent difference of SCE means in middle-aged and old groups. SCE counts were significantly lower for chromosomes A1 and C in old subjects. In some families of long-livers, SCE values of parents aged 80 years and older were lower than those of children. Probably, results of the above cultural studies are at variance with our findings because of the old ages of our subjects (80 years and over).

SCE findings in patients with rare autosomal recessive diseases (xeroderma pigmentosum, Bloom's syndrome, Fanconi's anemia, ataxia-telangiectasia, Louis-Bar syndrome) associated with premature aging and shorter lifetime are as controversial as in vitro studies (Pandata et al., 1995). Lymphocyte SCE rates in homozygous patients with Bloom's and Werner's syndromes have been reported to be 10- to 25-fold higher than control values (Chaganti et al., 1974; Bartram et al., 1976; Andriadze et al., 1986). However, study and control SCE levels proved similar in the usual cultures of lymphocytes or skin fibroblasts from patients with xeroderma pigmentosum (Wolff et al., 1975; Kato and Stich, 1976), Fanconi's anemia (Chaganti et al., 1974; Latt et al., 1975; Sperling et al., 1975; Louis-Bar syndrome (Chaganti et al., 1974; Galloway and Evans, 1975; Wolff, 1977), and Down's syndrome (Yu and Borgaonkan, 1977).

Most of the studies indicate nonrandom patterns of SCE locations over chromosome lengths, although evidence on hetero- and euchromatin incidence of sister chromatid exchanges is discrepant. An early work suggested higher SCE rates in the heterochromatin regions. Thus, the SCE BrDU-revealed incidence per unit of length of constitutive heterochromatin in *Microtus agrestis* X chromosomes was found to be about twofold higher than in euchromatin (Natarajan and Klasterska, 1975). Evidence for predominance of SCE in centromeric constitutive heterochromatin was obtained in labeled chromosomes of Chinese hamster (Marin and Prescott, 1964), rat kangaroo *Dipotomys ordii* (Gibson and Prescott, 1972), mouse (Holmquist and Comings, 1975), and human (Tice et al., 1975) chromosomes. Great percentages of sister chromatid exchanges were documented in G-banded regions or at the euchromatin-heterochromatin junction (Carrano and Wolff, 1975; Smith and Evans, 1976; Crossen et al., 1977). However, more recent studies suggest a higher frequency of SCE in chromosome euchromatin. The frequency was well below theoretical estimates (per unit of DNA) in constitutive heterochromatin of BrDU-treated human chromosomes (Galloway and Evans, 1975; Sperling et al., 1975; Ueda et al., 1976), animal sex chromosomes (Hsu and Pathak, 1976; Carrano and Wolff, 1975), and *Allium cepa* plant chromosomes (Schwartzman and Cortes, 1977). Hoo and Parslow (1979) have employed the technique for SCE identification in eu- and heterochromatin of human metaphase chromosomes; the SCE incidence in euchromatin was threefold higher than in heterochromatin, and SCE in the early replicating X chromosome were more common than in the late replicating, inactivated X chromosome.

Evidence on age-related incidence and locations of SCE in peripheral lymphocyte chromosomes suggests that SCE rates are lower in telomeric and centromeric (heterochromatin) regions of chromosomes 1, 2 and 3 and higher in the medial arms of these chromosomes. Our studies confirm that SCE prevail in chromosomal euchromatin and tend to decrease at 80 years or later (the time of excessive genomic heterochromatinization) (Lezhava and Chitashvili, 1982; Lezhava, 1980, 1984b, 1987, 2001a).

Conclusion

Studies of unscheduled DNA synthesis in peripheral lymphocytes exposed to UV irradiation (10 to 15 J/mm^2) showed it was substantially impaired in humans of 80 to 90 years of age, as compared to middle-aged (20- to 43-year-old) individuals. The evaluation of sister chromatid exchanges at ages of 80 and over revealed that single-metaphase SCE counts tended to be lower than middle-age values. A significant decrease in the SCE incidence is observed in A1 and C chromosomes of old subjects ($P < 0.05$). In some families, lymphocyte SCE counts of parents aged 80 or older were below those found in children. The analysis of SCE distribution over lengths of lymphocyte chromosomes A1, A2 and A3 demonstrated that in senile and control groups SCE were most abundant on medial arms and abruptly subsided in telomeric and centromeric directions, thus indicating a lower exchange incidence in heterochromatic regions.

CHAPTER IV

CHROMOSOME MODIFICATION

4.1. FUNCTIONAL DIFFERENTIATION OF CHROMATIN

Chromosome function in eukaryotes is related to structural and functional differentiation of chromosomes into condensed and decondensed regions. Condensed chromatin has three major features: (1) the bulk of it is made of heterochromatin, as was demonstrated in *Microtus agrestis*; (2) condensed chromatin contains abundant satellite DNA, as was proved by in situ hybridization studies, and this makes condensed chromatin equivalent to constitutive heterochromatin; (3) nuclear membrane-adjacent, nucleolus-associated condensed chromatin in electron microscopic sections of interphase cells betrays heterochromatic regions of chromatin (Hsu, 1975; Mc Dowell et al.,1999; Mantovani et al.,2000).

Studies of spermatogenesis in *Carixidae* showed that autosomes in sperm cells undergo complex meiotic transformations, while sex chromosomes (XY) long remain condensed (heteropyknotic) and lie in the nucleus as compact bodies. The sex chromosomes are also condensed in spermatogonia and spermatids, while autosomes unspiral to become fine strands (Prokofyeva-Belgovskaya, 1986). Heteropyknosis was identified not only in developing sex cells but also in chromosomes or chromosomal regions of somatic cells (Heitz, 1929, 1931). Heitz stated that both a sex chromosome and an autosome may be heteropyknotic.

The heteropyknotic chromosomes or chromosomal regions that do not decondense in the interphase nucleus and are heavily stainable by common dyes and genetically inert were named by Heitz heterochromatin; its antithesis, euchromatin, does not condense in the interphase nucleus, and is poorly stainable and genetically functional.

Heterochromatin regions are differentiated continuous chromonema bundles that run throughout the chromosome. Chromomeres of the heterochromatin and euchromatin regions have similar sizes, staining ability, distances between them and incidence per unit of chromosome length. Electron microscopy shows no difference in the fine structure of the eu- and heterochromatin regions (Ris, 1957).

Specific features of heterochromatin, making it different from euchromatin, are genetic inactivity, control of signs, heteropyknosis, intense staining, late replication, production of interphase chromocenters, instability, control of meiotic chiasm production, association with the nucleolus, susceptibility to breaks, and the presence of a reiterated DNA nucleotide sequence (Kurnit, 1979; John and Miklos, 1979; Prokofyeva-Belgovskaya, 1986;Cremer et al.,2000).

Krzanowska and Bilinska (2000) showed that centromere aggregation in the Sertoli cell progresses throughout the life of a male in a strain specific manner. Heterochromatin is further classified into constitutive, or permanently present in both homologous chromosomes throughout the cell cycle, and facultative, or identifiable in one of homologous chromosomes (Heitz, 1929).

Constitutive heterochromatin is organ- and species-specific. Initial studies of constitutive heterochromatin using differential C-banding of human chromosomes (Arrighi and Hsu, 1971) revealed heterochromatin variability which was defined as polymorphism (Craig-Holms and Show, 1971). C- and Q-banding studies have detected constitutive heterochromatin polymorphisms in chromosomes 1, 3, 4, 9, 13, 14, 15, 16, 21, 22 and in the Y chromosome (Buckton et al., 1976; Haaf and Schmid, 2000). Heterochromatin polymorphism is characteristic for individual genetically controlled chromosomes; it is inherited as a Mendelian trait and is almost uniformly observed in afflicted and phenotypically normal individuals (Buckton et al., 1976; Van Dyke, 1977; Podugolnikova and Korostelev, 1980; Howard et al., 1996).

Facultative heterochromatin occurs as a condensed (heterochromatinized) or decondensed (euchromatic) version; it does not have a paired pattern on homologous chromosomes or species specificity, and its location in the interphase nucleus is not specific. Prokofyeva-Belgovskaya (1986) indicated that the term facultative heterochromatin is valid for only one of the two female mammalian X chromosomes since it may not be appropriate for all homozygous inversions and translocations; when an euchromatic site in both homologous chromosomes is adjacent to a heterochromatic region, the former sustains heterochromatinization to become the facultative heterochromatin in both homologs (Brown, 1966). According to Prokofyeva-Belgovskaya, this is a good reason for dropping the term 'facultative heterochromatin' in favor of 'inactivated euchromatin' as a more descriptive definition. When inactivated euchromatin is heterochromatinized, it behaves as constitutive heterochromatin, i.e. it is condensed in the interphase nucleus and replicates late. Chromosome cycle studies of Prokofyeva-Belgovskaya (1986) and the general heterochromatinization concept make feasible the euchromatinization of condensed inert heterochromatin regions and heterochromatinization of active euchromatin regions. The arabidopsis mutation causes a striking decondensation of centromeric heterochromatin, a re-distribution of the remaining methylation of DNA, and a drastic change in the pattern of histone modification (Probst et al., 2003).

Heterochromatinization and euchromatinization are physiological conditions that may be long maintained at any chromosomal region regardless of cyclical conditions of other regions. Eu- and heterochromatinization are influenced by the nuclear chromosome composition, sex, parental age, temperature, and centromere vicinity (Prokofyeva-Belgovskaya, 1986; Carvalho et al., 2001).

Heterochromatinization restricts genetic activity during white and red blood cell maturation (Harbers and Sandritter, 1973), malignant neoplastic growth and in vivo and in vitro aging (Harbers and Sandritter, 1973; Ronne, 1980; Lezhava, 1980, 1984b; Lezhava,

2001 a, b). A change in the condensation (heterochromatinization) rate of a satellite stalk, which is made of heterochromatic segments of human acrocentric chromosomes, is thought to reflect a change in ribosomal cistron synthesis (Orey, 1974; Lezhava and Dvalishvili, 1992; Lezhava, 2001b).

The allocyclical Y chromosome of *Drosophila melanogaster* illustrates the relationship between the cyclic chromatin structure and gene activity. The Y chromosome is heterochromatinized in all *Dmelanogaster* tissues, except for the testes. However, DNA of a few Y chromomeres in primary sperm cells unwinds to make large loops, and high-rate RNA synthesis begins on these. The loops disappear when the synthesis is suppressed with actinomycin D (Khesin and Leibovich, 1976). It follows from the evidence outlined above that condensation is not a static but a dynamic process incurring structural and functional adaptations.

When chromosome rearrangements bring the euchromatin of chromosomes with certain loci close to heterochromatin, genes for somatic cell traits are activated. This change was defined as a position phenomenon (Lewis, 1950). Effects of X chromosome proximity to heterochromatic regions were examined in mosaic W^{mt4} and W^{mt5} lines of *D.melanogaster* (Prokofyeva-Belgovskaya, 1986). Locus 302, an active mediator of facet and testes pigmentation, exhibited scarce heterochromatin in 26.6 percent of salivary cell nuclei in W^{mt4} larvae, and flies had pigmented eyes, that is, the dominant trait (dark mottled) prevailed over the recessive one. In W^{mt5} larvae, locus 302 was adjacent to most of the chromosomal inert locus 4, and 81.2 percent of the nuclei were prominently heterochromatic. The eyes of these flies are made mainly of nonpigmented facets; therefore, the recessive trait (light mottled) was stronger in this line. These studies showed that traits controlled by a locus may develop a recessive pattern when the locus changes its euchromatic status for a persistently heterochromatic one (the position phenomenon).

When heterochromatin-located genes (responsible for ribosomal RNA synthesis) are transferred to euchromatin by a chromosome inversion, they are inactivated; as a result, *Drosophila* males with an XO genotype die at early developmental phases (Baker, 1974). The position phenomenon remains obscure. A hypothetical mechanism for it is inactivation of euchromatin genes that are transferred to heterochromatin by specialized proteins (histones?); the proteins may bind to heterochromatin satellite DNA and, probably, euchromatin DNA. This repression occurs after DNA replication early in development and later interferes with a normal gene function (transcription) and replication. Interaction with condensing proteins may explain the mosaicism of a single species displays in different cells. These proteins bind to DNA sites of various lengths. This process must be controlled by specialized genes. The total regulation of inactivation may involve modifier genes for mosaicism (Birstein, 1976; Howard, 1996).

The persistent gene inactivation exemplified by the position phenomenon may be a major mechanism underlying cell and tissue differentiation and, hence, a vital ontogenetic factor. In the presence of the position phenomenon, the structural modification of gene action recurs in cell generations. The phenomenon is similar to events induced by sex chromosome inactivation. Single X-chromosome inactivation in female mammalian cells is a facultative heterochromatinization case which has been elucidated in full detail.

The chromatin mass was originally identified in interphase nuclei of feline motor neurons (Barr and Bertram, 1949); it was later found in mammalian and human female tissues and defined as sex chromatin (Moor and Barr, 1954). Sex chromatin was thought to originate from heterochromatic sites of two X chromosomes. It was postulated by Stewart (1961) that a sex chromatin-producing factor locates in an X-chromosome long arm, i.e. a selected long-arm site of one X chromosome is heterochromatinized in human interphase nuclei. Confirmatory evidence was obtained in the studies of German (1962). One of the two X chromosomes replicated much later and other chromosomes were autosomal in cultures of female[3H]-thymidine-labeled lymphocytes. Heterochromatinization of the medial long arm of the human X chromosome was documented in the trophoblasts of 12-day embryos (Park, 1957).

Based on her *Drosophila* studies, Prokofyeva-Belgovskaya (1986) proposed a hypothesis that the heteropyknotic status of activated chromosome regions results in gene inactivation at a heterochromatic interkinetic nuclear region. This mechanism was redescribed by Lyon in 1962 and became known as lyonization. The Lyon hypothesis states that (1) a heterochromatic X chromosome is genetically inactivated in somatic cell nuclei; (2) the inactivation is random: it involves either a father- or mother-donated X chromosome; (3) inactivation-heterochromatinization occurs at an early embryogenetic stage.

A shared feature of two condensed chromatins (constitutive and facultative) is the absence of transcription and replication in the late S-period. What makes them different is DNA abundance in, and differential Giemsa staining of constitutive heterochromatin. Heterochromatinization presents itself as a physical change of chromosomes during condensation and a chemical change of chromosome proteins.

Percentages of highly repetitive DNA sequences (proportions of satellite to total DNA) may change with age (Prashad and Cutler, 1976; Suzuki et al., 2002). Mouse hepatic cells showed 7-8 percent of satellite DNA at ages of 10 to 300 days and 12-13 percent at 300 to 600 days (Prashad and Cutler, 1976). Studies in 6 strains of human diploid fibroblast cultures (Shmookler et al., 1983; Riabewol et al., 1985) have demonstrated late-passage amplification of 'inter-Alu DNA', which is a genomic DNA sequence interspersed between clusters of Alu repeats. This DNA was found to be extrachromosomal and covalently closed. Lymphocytes from 8 young (21 to 31 years) and 9 old (61 to 91 years) subjects were evaluated by these authors for the extrachromosomal inter-Alu DNA; it was present in the old donors' lymphocytes and absent in young lymphocytes. The amplified circular inter-Alu DNA molecules occurred mostly in B lymphocytes. Alu/Alu clusters and transposons were presumed to be similar in prokaryotes and eukaryotes. In addition, lower contents of tandem DNA repeats, C-Ha-ras-1 proto-oncogene amplification and higher contents of specific mRNAs and proteins have been reported in the genomes of aging human diploid fibroblasts (Goldstein et al., 1985).

4.1.1. Variability of Condensed Chromatin

This chapter presents evidence on condensed chromatin contents in blast-transformed lymphocytes from subjects of 25 to 90 years of age (Lezhava et al., 1979a; Lezhava,1999).

The electron microscopic study examined 309 blast-transformed interphase lymphocytes from 16 subjects of both sexes; 9 were apparently normal individuals aged 80 to 90 years who donated 200 interphase cells and 7 were aged 25 to 52 years (100 interphase cells).

The lymphocytes were cultured as recommended by Moorhead et al. (1960). A 72-hour cell suspension was placed into 2 percent phosphate-buffered (pH 7.4) glutaraldehyde solution for 2 hours at room temperature. The sediment was washed with a cool phosphate buffer (pH 7.4) and centrifuged at 800 g, and the cell suspension was fixed for 1 hour in 2 percent osmic acid on a phosphate buffer (pH 7.2). The suspension was washed with the phosphate buffer for 5 min to remove osmic acid and then embedded in araldite.

Ultramicrotome lymphoblast sections were examined and photographed in a UEMV-100 V electron microscope at x 10,000. Selective blast-transformed cells had chromatin lumps of comparable sizes (although G2 interphase nuclei had greater chromatin masses than G1 ones) (Lezhava et. al., 1979a). Percentages of condensed chromatin were calculated on lymphoblast photographs using the random-step morphometric grid of Stefanov (1974). A chart for computing standard errors and confidence limits was used and yielded a confidence level of 95 percent ($P > 0.05$).

These studies showed that a great percentage of cells were blast transformed (45.8 percent). The lymphoblasts were enlarged, variable in shapes and had round or ellipsoidal nuclei. Their nucleoli were acentric or membrane-adjacent. Lymphoblast chromatin occurred as finely granulated lumps. Applications of the morphometric grids to the electron microscopic cell images showed (Fig.4.1.1-1) that condensed chromatin values at ages of 80 to 99 years (67.8 ± 0.6 percent) significantly exceeded those of 25 to 52-year-old subjects (62.3 ± 0.9 percent).

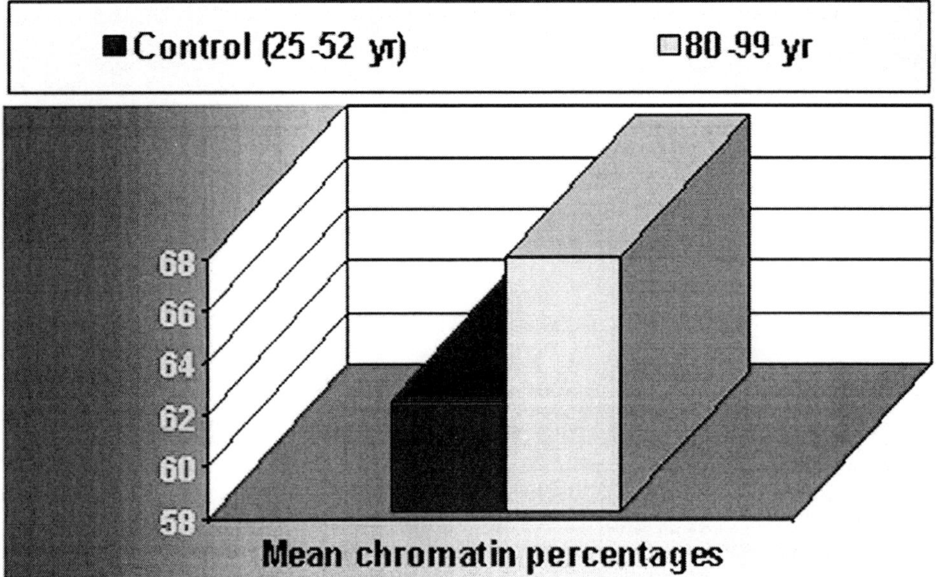

Fig.4.1.1-1. Interphase percentages of condensed chromatin in peripheral lymphocyte cultures by electron microscopy.

Moreover, condensed chromatin percentages in the old-age group varied in the extensive range from 58.9 ± 1.6 to 80.9 ± 0.8 percent (versus 57.7 ± 1.7 to 66.8 ± 1.4 percent in middle age).

Some old subjects had condensed chromatin percentages below the upper limit of a middle-age range (58.9 ± 1.6 to 68.6 ± 0.7 percent), while another subgroup aged over 80 tended to have progressive heterochromatinization (70.0 ± 1.9 to 80.9 ± 0.8 percent).

Our results indicate that aging of differentiated cells is associated with excessive heterochromatinization, and persons older than 80 years may fall into two subpopulations with different levels of chromosome thread heterochromatinization.

4.2. HEAT ABSORPTION OF CONDENSED CHROMAIN

Our study has examined chromatin condensation levels in lymphocytes from very senile subjects using differential scanning microcalorimetry (DSM) (Monaselidze et al., 1981). This technique allows measuring chromatin in milligram quantities. The heat process in whole chromatin is known to produce distinctive easily reproducible heat absorption peaks on heat capacity curves (Monaselidze et al., 1981; Lezhava et al., 1993). It has been found that ribonucleoprotein (RNP)-complex denaturation occurs in the temperature range of 40° to 65°C (maximum, 57°C). The temperature range of 60° to 90°C has two intensity maxima, *Td* ~ 66° and *Td* ~ 80°, corresponding to decondensed and condensed chromatin (Touchette and Cole, 1985). The DSM technique has a sensitivity of 10^{-7}cal/s and affords cell measurements in a 0.03 to 0.4-ml volume and in a temperature range of 20° to 150°C; heat rates vary from 1 to 0.05 kCal/h, and the accuracy of RNP complex and chromatin *Td* determination is ± 1°C. Chromatin was evaluated in nonstimulated lymphocytes from 4 donors of 20 to 35 years of age and 6 donors aged 77 to 82 years (curves 1 and 4) and phytohemagglutinin-stimulated (PHA, Difco P) lymphocytes from 4 donors aged 20 to 35 years and 6 donors aged 77 to 82 years(curves 2 and 3). The lymphocytes were cultured as usual.

It can be seen from Figure 4.2-1 (curves 2) that 40-h PHA stimulation of the cultures significantly influenced chromatin modification. Heat redistribution between the heat absorption peaks was obvious. In particular, the decondensed chromatin (euchromatin) intensity peak increased and the condensed chromatin (heterochromatin) high-temperature peak decreased twice; the peak maxima did not shift along the temperature scale, unlike those of nonstimulated lymphocyte chromatin (curve 1). This means that a condensed chromatin fraction was activated, and it melted with the decondensed fraction.

A different pattern was seen in aged subjects. Curve 3 demonstrates that heat redistribution between the peaks was paralleled by shifts along the temperature scale. The high-temperature peak shifted upward by ~3°C, and the low-temperature peak moved downward by ~4°C. Curve 4 indicates that chromatin from the unstimulated lymphocytes was more condensed in comparison with that in young donors – shift of peak maxima was observed in chromatin of old individuals (Fig.4.2-1). These prominent changes in chromatin stability indicated transformation of condensed chromatin (heterochromatin) into decondensed chromatin (euchromatin).

The chromatin peak behavior described above showed progressive lymphoblast chromosome heterochromatinization in old individuals confirming previously reported evidence (Lezhava, 1980, 1984b; Lezhava and Dvalishvili, 1992; Lezhava et al.,1993; Lezhava, 2001b).

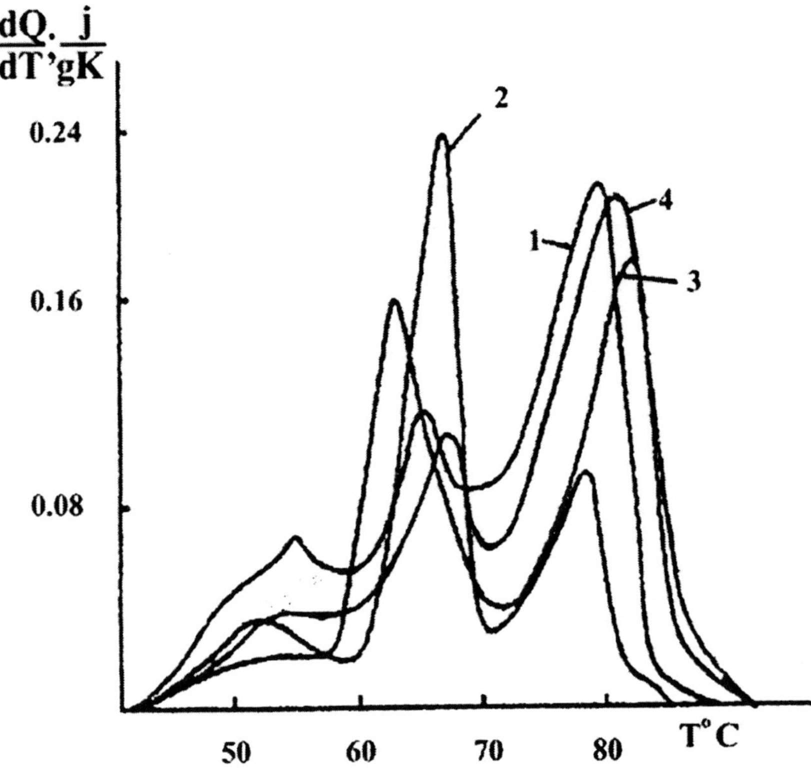

Fig. 4.2-1. Heat absorption curves of human lymphocytes suspension (recalculation per gram of dry substance). 1 - nonstimulated lymphocytes from young individuals; 2 - PHA- stimulated lymphocytes from young individuals; 3 - PHA- stimulated lymphocytes from old individuals; 4 - nonstimulated lymphocytes from old individuals.

The lymphocytes generally contain inactive condensed chromatin, and 99.9 percent are nondividing cells (Evans and Norman, 1968). Lymphocyte division begins at 24 hours following PHA stimulation, (Cooper et al., 1961). The process is associated with structural chromatin modification. Lymphocyte transformation into lymphoblasts corresponds to gene activation. Twenty-four hours after PHA stimulation condensation decline and euchromatin expansion were observed, indicating heterochromatinization reversal.

The chromatin peak behavior described above showed progressive lymphoblast chromosome heterochromatinization in persons of 80 years and older ages, confirming previously reported evidence (Brown, 1966; Kanungo, 1984; Lezhava, 1980, 1999, 2001a).

4.3. Centromeric Heterochromain

The morphology of metaphase chromosomes has been microscopically evaluated in 24 subjects (Lezhava, 1977); selective pericentromeric staining by Unna's blue occurred in 6 subjects aged 81 to 114 years and was absent in the control group ranging in age from 13 to 34 years. Table 4.3-1 presents the rates of centromeric heterochromatin staining. C-band locations were similar to those seen after an alkaline or thermal pretreatment or staining with buffered Giemsa.

Stain locations varied in homologous, group-specific and intergroup autosomes. Large blocks of centromeric heterochromatin were common on homologous chromosomes A1qh (Table 4.3-1). In a percentage of metaphases, the heterochromatin-positive 1qh chromosomes displayed some packing impairment. Sizes and distribution of centromeric heterochromatin on the A1qh homologs varied in some metaphases of 6 studied individuals.

Of interest was a sample from a 114-year-old man whose A1qh showed dark-stained heterochromatin sites sized 1.5-fold greater than counterpart sites in other individuals' samples. However, intrahomolog variability was often related to sizes and the incidence of heterochromatin blocks (Fig. 4.3-1).

To find out whether or not the heterochromatin blocks resulted from DNA denaturation and successive renaturation (DNA is denaturated as a fixative is burnt; De la Chapell et al., 1971), another sample was obtained from that old man 10 months later. A proportion of metaphase chromosomes were prepared by fixative burning and another one by air-drying. A few chromosome preparations were exposed to hydrochloric acid hydrolysis. Unbuffered Unna-blue (pH 7.1) was used to stain the samples.

Assay results were similar to the previous findings: chromosome 1qh metaphases of the 114-year-old man were positive for the heterochromatin blocks in the absence of thermal shock-induced DNA denaturation. However, less metaphase had centromeric heterochromatin, and centromeric regions of the homologous chromosomes showed a lower staining. Centromeric heterochromatin staining remained dominating on one or both A1qh homologs.

4.3.1. Y-Chromosome Centromeric Heterochromatin in Longeval Families

The recurrence of constitutive heterochromatin sizes has been assessed in Y-chromosome C bands (Fig.4.3.1-1) in 3 families. Chromosome photographs were scanned in order to measure optical density distribution over chromosome surfaces. The data were processed using computer programs (Saralidze et al., 1981), and this provided dimensions and locations of various chromosome regions. The morphometric system has been designed by Nisanov et al. (1981), at the Institute of Molecular Biology, Moscow. The accuracy of the measurements depends on two factors: (1) precision with which chromosome, E- and C-band sizes are obtained; (2) a methodology for comparing the findings. This morphometric system offers considerable advantages over earlier approaches.

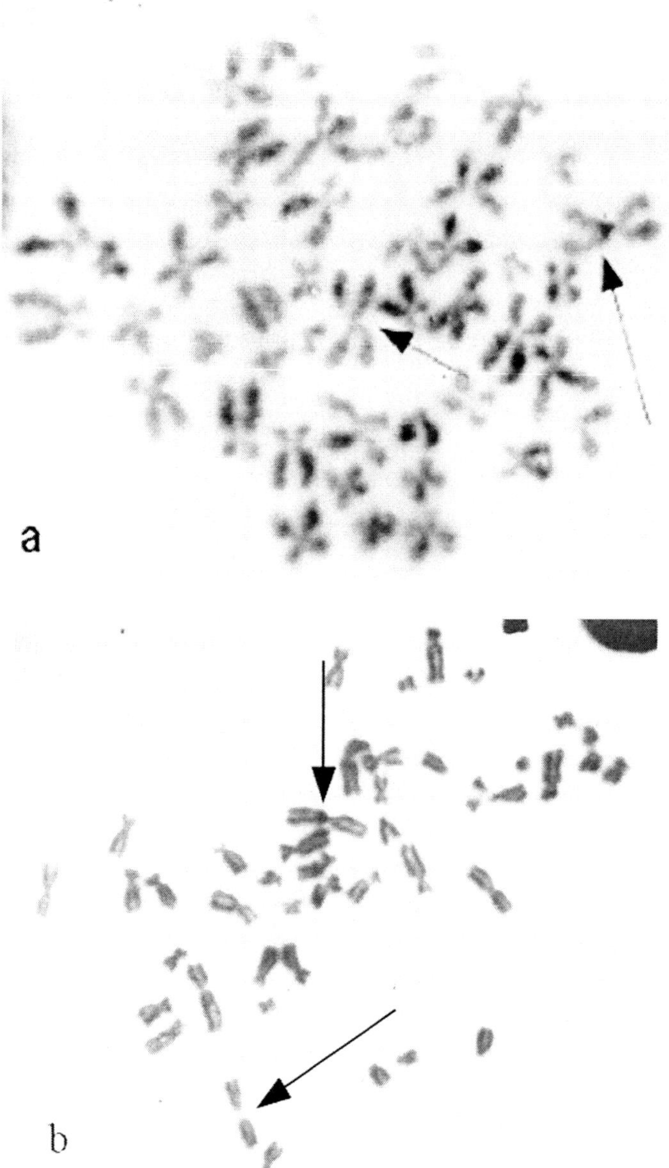

Fig. 4.3-1. Distribution of C - bands on one of the homologs of the A1 qh chromosomes without preparation pretreatment and unbuffered Unna blue staining. Arrows: 114-year-old man (a); 83-year-old man (b).

Heterochromatin (C-band) variants were evaluated using the Sumner technique (1972). Chromosome preparations were treated with 0.1 percent HCl solution for 30 min at 20° C, washed with three portions of distilled water and put into a Ba(OH)2 solution for 10 min (55°C). After the incubation the preparations were washed with distilled water and immediately placed into a 2 x SSC solution (60°C) for 1 h. Then they were washed in running water and stained with 5 percent Giemsa on a phosphate buffer (pH 6.8) during 1 h.

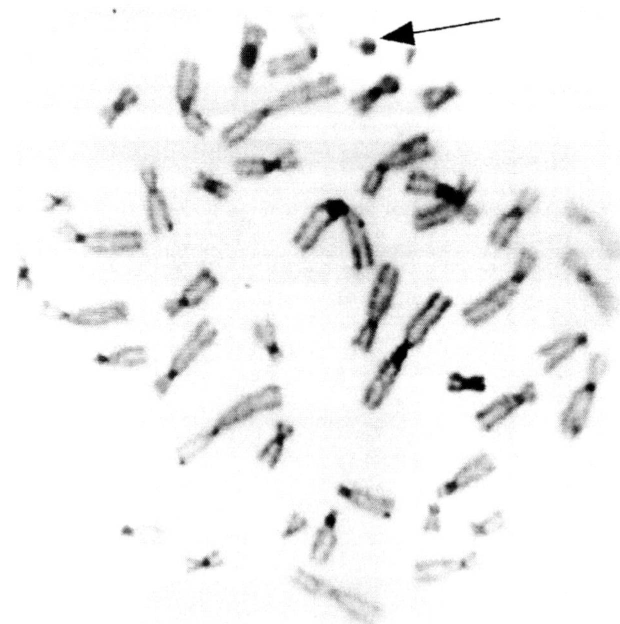

Fig. 4.3.1-1. Differentially C-stained metaphase of a senile man. Arrow: Y chromosome.

Table 4.3.1-1 presents morphometric Y chromosome C heterochromatin findings in normal family members. Both similarity and age-specific variability were found in sizes of E euchromatin and C heterochromatin bands.

Table 4.3.1-1. Comparative intrafamilial E- and C-band measurements (μm) of peripheral lymphocyte Y chromosome

Generation	I		II		III	
Families	**E**	**C**	**E**	**C**	**E**	**C**
	0,65	0,6	0,6	0,6	0,7	0,4
I	0,85	0,7	0,9	0,7	0,8	0,4
	0,9	0,7	0,95	1,0	0,85	0,5
	0,9	0,9	1,0	0,9	0,9	0,4
	88 yr		59 yr		30 yr	
	0,9	0,7	0,7	0,8	—	—
	0,9	0,8	0,8	0,7	—	—
II	0,9	0,8	0,9	0,7	—	—
	0,95	0,85	1,0	0,8	—	—
	1,0	0,8				
	86 yr		55 yr			
	0,7	0,6	0,85	0,8	—	—
	0,7	0,6	0,9	0,7	—	—
III	0,7	0,7	0,9	0,7	—	—
	0,75	0,6	1,05	1,0	—	—
	0,9	0,9	1,2	1,0	—	—
	83 yr		51 yr			

Note. E - euchromatin band; C-heterochromatin band

With a B-chromosome spiralization index lying in the range of 4 to 7 μm, the E-band sizes did not significantly differ within age groups of families 1 and 2. In family 1, a mean E-band size was 8.2 at 88 years, 8.6 at 59 years and 7.1 at 30 years. The difference was not significant. Likewise, the E-band sizes did not significantly vary in family 2 (9.3 in the 86-year-old father and 8.5 in his 55-year-old son). A difference of the E-band sizes was seen in family 3 (7.5 in the 83-year-old father and 9.4 in his 51-year-old son).

A dimensional variability of C heterochromatin bands occurred in family 3: the band size was 6.8 at 83 to 88 years and 8.4 at 51 to 59 years.

It appears that the variability of the findings was determined by different spiralization indices. Proportions of E to C band sizes were similar at ages of 83 to 88 years and 51 to 55 years. A sole exception was a 30-year-old individual in family 1: his Y-chromosome C band was shorter than that of his father and grandfather (Table 4.3.1-1).

Our findings suggest intact Y-chromosome E and C bands and their Mendelian pattern of inheritance in apparently normal individuals aged 83 to 88 years and 51 to 59 years.

4.4. Telomeric Heterochromatin

Telomere is the structure representing heterochromatin, which is located on both ends of individual chromosomes in eukaryotes. The DNA sequence of the telomere consists of Guanine-rich tandem repeat i.e., (TTAGGG) n, in man. Telomere protects the end of the chromosome from fusion or deletion and maintains the stability and replication of chromosome and is synthesized by telomerase, a ribonucleoprotein. Telomere reduction is observed with cell senescence and immortalization, both in vivo and in vitro. Thus, telomere is a "clock" which measures the life span of cells and its length is altered by cellular senescence and immortalization (Greider, 1991). The telomere hypothesis of cellular aging proposes that loss of telomeric DNA (TTAGGG) from human chromosomes may ultimately cause cell-cycle exit during replicative senescence. Lymphocytes have a limited replicative capacity and blood cells were previously shown to lose telomeric DNA during aging in vivo. It was determined: (a) accelerated telomere loss is associated with the premature immunosenescence of lymphocytes in individuals with Down's syndrome (DS) and (b) telomeric DNA is also lost during aging of lymphocytes in vitro (Greider, 1991; Ohmura and Oshimura, 1993).

The rate of telomere loss was calculated from the decrease in mean TRF length, as a function of donor age. DS patients showed a significantly higher rate of telomere loss with donor age (133 +/- 15 bp/year) compared with age-matched controls (41.0 +/- 7.7 bp/year) ($P<0.005$), suggesting that accelerated telomere loss is a biomarker of premature immunosenescence of DS patients and that it may play a role in this process. Telomere loss during aging in vitro was calculated for lymphocytes (from four normal individuals) grown in culture for 10-30 population doublings. The rate of telomere loss was approximately 120 bp/cell doubling, comparable to that seen in other somatic cells (Vaziri et al., 1993).

Telomeric DNA in the skin cells of 21 human individuals aged between 0 and 92 years was quantified by determining the length of the telomeric smear and the relative amount of TTAGGG repeat sequences. Both telomere length and quantity of telomeric repeat sequences were found to decrease significantly with age (Lindsey et al., 1991). Telomere shortening occurs more rapidly on human X chromosomes, which could contribute to the age-dependent reactivation of X chromosome loci (Campisi, 2000).

Data showing a decrease of about 50 base-pairs per generation in fibroblasts suggest that a full deletion event is 100 to 200 base-pairs. If cells senesce after 80 doublings in vitro, mean telomere length decreases to about 4,000 base-pairs, but one or more telomeres in each cell will lose significantly more telomeric DNA. A checkpoint for regulation of cell growth may be signalled at that point. Variation in telomere length predicted by the model is consistent with the abrupt decline in dividing cells at senescence. Variation in length of terminal restriction fragments is not fully explained by incomplete replication, suggesting significant interchromosomal variation in the length of telomeric or subtelomeric repeats. This analysis, together with assumptions allowing dominance of telomerase inactivation, suggests that telomere loss could explain cell cycle exit in human fibroblasts (Levy et al., 1992). Telomerase has an important role in the repair of G-rich DNA (Lansdorp, 2005).

A zinc-finger gene encoding a transcription factor that regulates hematopoiesis, MZF-1, is located at the extreme end of the q arm of human chromosome 19. Several lines of evidence indicate that MZF-1 lies less than 20 kb from the subtelomeric repeat region of 19q. Telomeres are known to degenerate as cells age; disruption of MZF-1 due to telomeric degeneration may play a role in the increased incidence of leukemia in the elderly (Hoffman et al., 1996). The frequency of telomeric repeats was variable in the ataxia telangiectasia cell lines (4.3 and 8.2 kb) compared to the normal cell lines (9.6 and 12 kb) and an inverse correlation between telomere length and chromosome end associations was observed. Both ataxia telangiectasia cell lines showed more robust telomerase activity than the normal cell lines, precluding defective enzymatic capacity as the basis for the chromosome end associations. It is possible that chromatin structure in the form of telomere-nuclear matrix interactions is variant in ataxia telangiectasia cells negatively influencing telomerase function and contributing to telomere associations (Pandata et al., 1995).

Cottliar et al. (2000) analysed chromosome instability on peripheral blood lymphocytes cultured from 7 untreated patients with chronic pancreatitis (CP) by assessing telomeric associations (TAS). Mean frequencies of TAS were significantly higher in CP patients (11.00 +/-2.37) compared to controls (1.00 +/-0.30) ($p<0,001$). Chromosomes preferantially involved in TAS were: 9, 20, 16 and 21, being the most affected arms: 9p, 20q, 16p, 9q and 21q. All these terminal bands were coincident with cancer breakpoints ($p<0.03$), two of them (40%) were specifically associated to pancreatic carcinoma rearrangements. Three bands (60%) were coincident with oncogene location. These results clearly indicate that CP patients exhibit chromosome instability, showing the presence of an unstable genome that could be related to the cancer development observed in this disease.

Since telomeres permit complete replication of eukaryotic chromosomes and protect their ends from recombination, Counter et al. (1992) have measured telomere length, telomerase activity and chromosome rearrangements in human cells before and after transformation with SV40 or Ad5. In all mortal populations, telomeres shortened by approximately 65 bp/generation during the lifespan of the cultures. When transformed cells reached crisis, the length of the telomeric TTAGGG repeats was only approximately 1.5 kbp and many dicentric chromosomes were observed. In immortal cells, telomere length and frequency of dicentric chromosomes stabilized after crisis. Telomerase activity was not detectable in control or extended lifespan populations but was present in immortal populations. These results suggest that chromosomes with short (TTAGGG)n tracts are recombinogenic, critically shortened telomeres may be incompatible with cell proliferation and stabilization of telomere length by telomerase may be required for immortalization.

End-to-end (pp, qq, pq) association of metaphase chromosomes was observed in extreme old age. Lymphocyte cultures were prepared by the usual method from the peripheral blood of 20 subjects including 4 women and 6 men aged from 80 to 114 years and 5 women and 5 men aged from 20 to 50 years (control group). In each case 20 metaphases were examined. Criteria for the determination of an association were as follows: each one of their arms is connected by fibre-like structures; if the distance between arms does not exceed the width of chromatid and the continuations of their axes cross. We studied the mean number of the end associations of type pp, qq and pq telomeres as well as the frequency of the mean number of telomere associations per cell. Our experimental data showed that the number of cells with end-to-end telomeres associations and the total frequency of aberrant telomeres were considerably increased at old age (0.63%; 1.84%) in comparison with those at middle age (0.44%; 1.02%) ($P<0.01$).

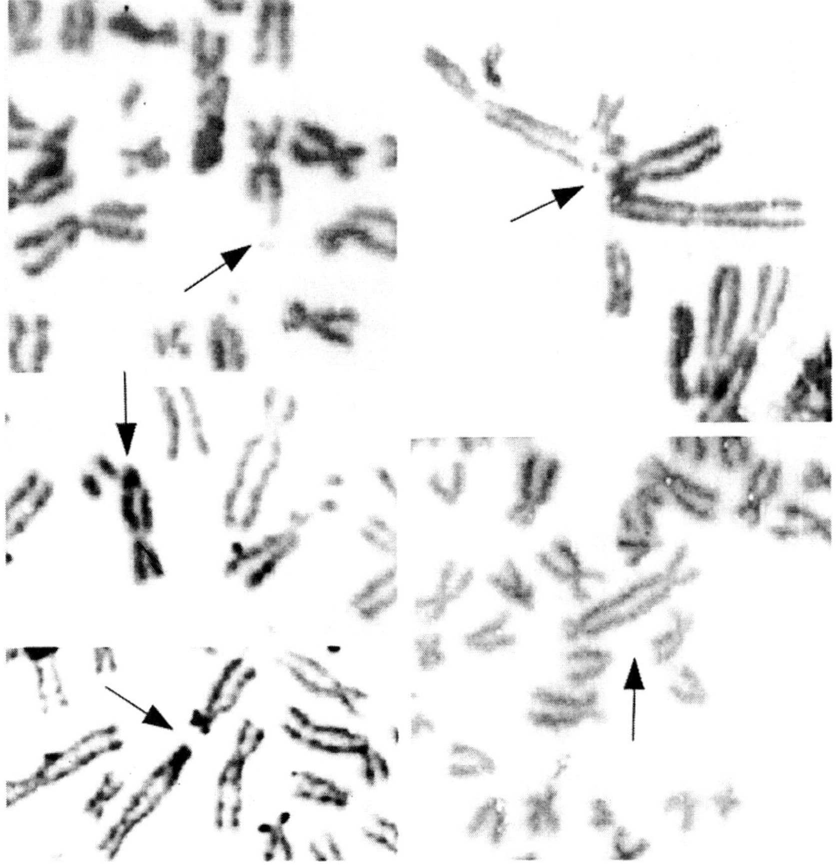

Fig. 4.4 – 1. Telomere aberrations and end-to-end associations of chromosomes from old individuals are shown by arrows.

The higher frequencies of chromosome end-to-end associations in extreme old age may be due to the loss of telomeric regions and form chromosome aberrations (Fig. 2.2.1-2 and 4.4-1). Telomere is a structure representing heterochromatin which consists of a Guanine rich tandem repetitive DNA sequence (TTAGGG)n in men. The telomere length reduction is observed with cell senescence, both in vivo and in vitro. Chromosomes with loss tracts are recombinogenic (Pandata et al., 1995). Loss of telomeric heterochromatin can be considered as a macromutation sharply changing chromosomal structure, though minimally affecting the phenotype. Loss of telomeres cannot be the reason for aging, but the result of aging processes, likely of chromatin in cyclops diminution (Akifiev and Grishanin, 1993; Akifiev et al., 2002).

A biological role of telomere loss, as a macromutation, indicates the exceptional plasticity of the human genome, which suffers fundamental changes during extreme old age.

Discussion

Our evidence indicates that in very old age cyclical properties of chromosome regions in lymphocytes vary with genome functioning. Small lymphocytes have dense nuclei and massive condensed chromatin lumps; 99.9 percent of these lymphocytes are nondividing and are in G_0,G_1 stage (Evans and Norman, 1968). In the study of Cooper et al. (1961) the small lymphocytes incorporated tritium-labeled thymidine, enlarged, transformed and initiated divisions at 24 hours following PHA stimulation. The radioactive label was detected in 50-70 percent of mononuclears after 70 hours of cultivation (Cooper et al., 1961; Drings and Sonnemann, 1974).

These processes are associated with structural chromatin modifications in lymphocyte nuclei. The transition of the small lymphocytes to lymphoblasts is a case of gene activation. After 24 hours of PHA stimulation, condensed chromatin decreases and euchromatin elevates in the cells (Drings and Sonnemann, 1974); deheterochromatinization begins early in the transformation, but it is demonstrable at light and electron microscopy only on 24-hr incubation of PHA-stimulated lymphocytes.

We have found rearrangements between heat absorption peaks. Forty-hour PHA stimulation of lymphocytes resulted in a double increase of the decondensed chromatin peak and a double decrease of the high-temperature (condensed chromatin) peak. The latter shifted to higher temperatures by about 3° C in very old age, indicating progressive heterochromatinization in persons of 77 years or older (Lezhava, 1984b; Lezhava and Dvalishvili, 1992). These data correspond to the results obtained by us. Electron microscopic studies of blast-transformed lymphocytes confirm that aging is associated with progressive heterochromatinization of dividing and senescent differentiated cells (Prokofyeva-Belgovskaya, 1986; Panno and Nair, 1984; Kanungo, 1984; Suzuki et al., 2002).

Haaf and Schmid (2000) propose that both DNA hypomethylation and steric alterations in chromosomal DNA may interfere with the binding of specific proteins or multi-protein complexes that are required for chromosome condensation. DNA repetitions may provoke heterochromatinization (Fourel et al., 2002)

Further, we have identified two subpopulations with different rates of chromosome thread heterochromatinization at 80 years or older. This provides support to the concept that humans make up two subpopulations: one prone to diseases and an early elimination of individuals and the other having a genetic potential for longevity which environmental factors may or may not allow to be fulfilled (Kuznetsova and Zaritskaya, 1986).

Centromeric heterochromatin segments detected on the 1qh chromosome of old individuals and electron microscopic evidence for the aging-related increase in lymphoblast condensed chromatin levels suggest a contribution of functional variability of selected chromosomal regions (including centromeric ones). Partial experimental confirmation of this has been provided by Prashad and Cutler (1976); they found satellite DNA to increase in murine hepatic cells with aging. Similarly, nuclear percentages of 'inactive' hepatic cell chromatin were higher and those of 'active' chromatin were lower in old rats, as compared to sexually mature rats (Mozzhukhina, 1984).

The variability of C-band constitutive heterochromatin accounts for the problem of saving chromosome polymorphism in dividing cells and identifying the recurrence of variant constitutive heterochromatin.

Variant C-band heterochromatin of chromosomes 1, 9 and 16 has been reported to be stable in cultures maintained for many years and in primary cultures of different tissues of an individual (Hoehn et al., 1977; Amchenkova et al., 1981). By contrast, recurrence patterns of C-band chromatin are controversial. Variant constitutive heterochromatin has been found to be altered by recurrence (Craig-Holmes et al., 1975). A large heterochromatin C band has been identified on one chromosome 9 in a patient with trisomy 9 whose parents were negative for that band (Seabright et al., 1976). A large satellite made of constitutive heterochromatin has been documented on chromosome 22 in a girl with congenital malformations and mental retardation, but not in her parents (Nakagome et al., 1977). Quantitative evaluation of C heterochromatin has identified different sizes of chromosome-9 C band in a father and his child (Balicek et al., 1978; Simi and Tursi, 1982).

It is notable that a statistically significant enlargement of chromosome 9 C-heterochromatin has been identified in women of 80 years and older (Kuznetsova and Zaritskaya, 1986). In a Bulgarian population, excessively large Y-chromosome variant constitutive heterochromatin occurred in 2.5 percent of 41 clinically normal newborns and in 17 percent of men over 60 (Tsoneva et al., 1980). On the other hand, familial recurrence of chromosomes 1, 9, 16 and Y C-bands was not associated with constitutive heterochromatin variability in posterity (Carnavale et al., 1976; Beltran et al., 1979).

It is obvious that the most common human Y chromosome polymorphism occurs in constitutive heterochromatin, which is made brightly fluorescent by Q-banding and prominent by C-banding. A lack or enlargement of the Y-chromosome C-band produces no phenotypical abnormalities. No relation has been found between physical or mental anomalies and C-band sizes (Paris Conference, 1971; Akesson and Wahlstrom, 1977).

Y-chromosome euchromatin to heterochromatin ratios proved similar in fathers and male babies: $1.14 \pm 0.03/1.09 \pm 0.03$ in fathers' venous blood, $0.99 \pm 0.01/0.94 \pm 0.02$ in sons' umbilical blood (Beltran et al., 1979).

Our comparative studies of Y-chromosomes with large C-bands in three families have not revealed E- and C-chromosome band variability at ages of 83 to 88 and 51 to 59 years.

The results of studies in senile individuals are consistent with reported evidence for stability and Mendelian inheritance of variant C heterochromatin.

The rate of telomer loss was calculated from the decrease in mean telomere length. Loss of telomeric regions cannot be the reason of aging, but is the result of aging processes.

CONCLUSION

Chromatin studies in very old individuals have yielded a confirmation of condensation variability. Differential scanning microcalorimetry identified a heat absorption peak indicative of heterochromatinization in old age. Electron microscopic evaluation of blast-transformed lymphocytes showed that condensed chromatin percentages in individuals of 80 years and older were above a middle-age range. In addition, karyotype analysis in lymphocyte cultures of apparently normal individuals of 80 to 114 years of age demonstrated that in a number of the subjects the A1qh chromosome was positive for centromeric heterochromatin in the absence of alkaline or thermal pretreatment by Unna staining (pH 8.9). However, the scanning and morphometric studies revealed no variability of Y-chromosome C-bands in 83 to 88 year-old fathers and 51 to 59 year-old sons. The results obtained indicate that chromosome telomeres change during aging.

CHAPTER V

ACROCENTRIC CHROMOSOMES

5.1. NUCLEOLAR ORGANIZER REGIONS

5.1.1. The Short Arm of the Human Acrocentric Chromosome: Morphology

The metaphase nucleolar organizer regions (NOR) are chromosomal regions in which ribosomal genes are encoded. The structural component of NORs is mainly constituted by 2 multifunctional proteins: the C23 or nucleolin, 110 kd and the B23 or nucleophosmin, 39 kd. NOR-associated proteins, whose abundance, turnover and localization are closely linked to ribosomal gene activity, are usually called "AgNOR" because of their argyrophilic properties (Cannavo et al., 2001).

The short arm of each of the ten human acrocentric chromosomes (pairs 13, 14, 15, 21 and 22) is classically made up of three regions: the centromere-adjacent short segment, the strand designed to be a secondary constriction, and the small compact body (satellite) which contains the short arm - telomere .

The acrocentric satellite stalk exhibits extensive phenotypical variability, which is related to labile chromosome spiralization (Ferguson-Smith and Handmaker, 1961; Bishun, 1966). Variations in the length of satellite stalk and the size of the latter are associated with a change in spirilization and, hence, chromosome functioning, depending on genotypical medium (Prokofyeva-Belgovskaya, 1986). Some individuals were found to have 'giant satellites' with Mendelian inheritance (Tjio et al., 1960) and double satellites, which were a large (proximal) and a small (distal) satellites on each satellite stalk. D group chromosome double satellites were described in a patient with multiple malformations (Groushy et al., 1964) and in a patient with Klinefelter's syndrome (Stahl et al., 1966). G group chromosome double satellites were identified in a hypospadia patient (Turpin and Lejeune, 1965) and in the Stein-Leventhal syndrome (Lezhava, 1999). The double satellites may recur in several generations and stimulate acrocentric chromosome nondisjunction (Stahl et al., 1966, Lezhava, 1999).

The inheritance of Mendelian fission was presumed for the satellites from the time of their identification (Ellis and Penrose, 1960). This hypothesis was filled with substance as satellite morphology and function were examined in human monozygotic and dizygotic twins. Studies in 17 twins (Engmann, 1967, 1972) demonstrated that both the incidence and morphology of the satellites and stalks were shared by the monozygotic twins; the ratio of D chromosome to G chromosome satellites was constant in most monozygotic twins and variable in dizygotic ones. The interesting implications of these findings were inheritance of the satellites, variability of satellite counts within Mendelian fission and, finally, availability of a simple test to identify monozygotic and dizygotic twins. Ford and Woolam (1967) showed that the ratio of D to G chromosome satellite incidence is individual-specific and thus provides an important descriptor of individual karyotypes. In addition, chromosome luminescence (Q-banding) studies found that the presence or absence of acrocentric satellites is an individual feature (Schmid et al., 1974).

Human acrocentric chromosomes with prolonged satellite stalks were reported in health and disease (Lezhava, 1966; Ford and Woolam, 1967). Acrocentrics with prolonged satellites have a high incidence in the population. Relatively long satellite stalks on one or more acrocentric chromosomes were identified in 29 of 100 apparently healthy subjects (Engman, 1972).

An example of chromosome changes is variability of acridine orange "reverse" banding and short-arm sizes of acrocentrics seen in 25 clinically normal subjects (Verma and Lubs, 1975).

It was demonstrated with the use of in situ hybridization that short arms of human acrocentric chromosomes contain genes encoding 18s and 28s ribosomal RNA synthesis (Henderson et al., 1972; Evans et al., 1974). Differential Ag-SAT staining of nucleolar organizers indicated that the human genes for 18s and 28s RNA locate only in acrocentric secondary constrictions - stalks (Chicago Conference, 1966; Goodpasture et al., 1976). Mutant *Xenopus laevis* embryos were unable to form ribosomal RNA in the absence of the nucleolar organizer (Brown and Gurdon, 1964). rDNA levels may vary with individuals, and chromosomal rRNA levels may vary with homologous and nonhomologous acrocentrics in the same individual (Guanti and Petrinelli, 1974; Evans et al., 1974). Biochemical evaluation of RNA types in a definite nucleolus suggested that the production of all types could not be controlled by the nucleolar organizer (Stahl and Luciani, 1972). It was shown by Perry (1969) that 45s is encoded by a nucleolar organizer locus, while 5s RNA is controlled by genes that reside outside it. The chromosome A1 telomere has at least one locus encoding 5s RNA (Henderson et al., 1980). The chromosome A1 itself may be associated with acrocentrics (New Haven Conference, 1973). This range of evidence indicates that nucleolar constituents are controlled by different chromosomal regions.

The involvement of acrocentric chromosomes in the nucleolus formation was documented by numerous studies, suggesting that an abnormally high number of genes encode high-molecular weight ribosomal RNA in trisomies 21 and 13 (Bross and Krone, 1973; Dittes et al., 1974; Miller et al., 1977). Nucleolus formation-mediating satellite stalks are mostly heterochromatin regions (Stahl and Luciani, 1972; Evans et al., 1974). Nucleolus-associated heterochromatin contains ribosomal cistrones, or genes involved in a crucial cellular process - the synthesis of proteins for cell growth and chemical and structural renewal. Unexpectedly, the nucleolus also seems to play a role in nuclear export, sequestering regulatory molecules, modifying small RNAs, assembling ribonucleoprotein (RNP) and controlling aging (Olson at al., 2000).

The view that heterochromatin represents inactivated regions unable of RNA synthesis was challenged by Stahl and Luciani (1972); it was pointed out that heterochromatin should be considered within a dynamic context, with reference to its ability for structural and functional adaptation. Besides ribosomal sequences, nucleolar organizers contain clusters of multiple DNA repeats and AT-rich satellite DNA (Gosden et al., 1979). A study in *Acheta domesticus* oogonia and oocytes showed that RNA synthesis is inactive during the meiotic interphase, when chromatin is a compact heterochromatin mass. However, the synthesis is activated and 3H-uridine incorporation occurs during the pachytene and diplotene stages, when RNA strands bulge, making small puffs (Lima-de-Faria, 1969). Similar results were obtained in *Acheta domesticus* by Stahl and Luciani (1972) who found that heterochromatinization was increasingly greater at stages from the first-order oocyte to the leptotene and declined thereafter. In exploring the mechanism by which the active silent state of rRNA genes is inherited, Santoro et al. (2002) found that NoRC, a nucleolar remodeling complex, containing Snf2h (also called Smarca 5, SWI/SNF-related matrix-association actin-dependent regulator of chromatin subfamily a, member 5), represses rRNA transcription. NoRC mediates rRNA silence by recruiting methyltransferase and histone deacetylase activity to the r-DNA promotor, thus establishing structural characteristics of heterochromatin, such as DNA methylation, histon hypoacetylation and methylation of the Lus 9 residue of histone H3 (Santoro et al., 2002).

Studies of acrocentric D and G group chromosomes in metaphases of cultured lymphocytes in health and disease (Lezhava, 1966) and in subjects 80 years and older (Lezhava, 1999) have revealed a variety of phenotypical satellite versions. Seven variants of satellite regions were identified and defined according to their configurations as a classical satellite which is a rounded body on the satellite stalk (Fig. 5.1.1-1*a*); an antenna satellite, or unwound well-visualizable strand (Fig.5.1.1-1*b*); a puff satellite which is a site with poor banding resolution of a poorly outlined strand (Fig.5.1.1-1*c*); a ring-shaped satellite, or unwound distinct strands that make a circular figure of their free ends (Fig. 5.1.1-1*d*); a blended satellite produced by converging satellites (Fig.5.1.1-1*e*); a spheroid satellite detached from the short arm (Fig.5.1.1-1*f*); and an 'unsatellited' acrocentric chromosome: the satellite stalk is so short that it is difficult to differentiate from the distal short arm of the chromosome.

The acrocentric chromosomes were categorized into three groups according to the satellite morphology; the criteria were forms of connections between chromatid satellites (Fig.5.1.1-1.). The first groups were chromosomes with unrecognizable satellites. The second group included acrocentrics with no links between chromatid satellites. The third group included acrocentric chromosomes with approximated or blended satellites. Heliot et al. (2000) studied the different AgNOR staining patterns of metaphase chromosomes in human lymphocytes. Three predominant patterns could be distinguished: pair (47%), stik-like (28%) and unstained (18%) structures.

The intersatellite links in the third-group acrocentrics were presumed to stimulate chromatid nondisjunction in the 2nd meiotic anaphase or zygote mitosis. It is noteworthy that a relationship was reported between advanced maternal age and maternal G group acrocentric chromosome nondisjunction associated with infantile Down's syndrome (Penrose, 1962; Lange et al., 1975).

Starting from this point, we have evaluated variability of the satellite morphology in acrocentric chromosomes (the presence of blended or unblended chromatid satellites) in relation to sex and age. The study material was 922 metaphases of lymphocyte cultures derived from 36 individuals; 658 metaphases were from 21 individuals (8 women, 13 men) aged 80 to 114 years and 264 metaphases from 15 individuals (9 women, 6 men) aged 14 to 48 years. Metaphases with similar spiralization and satisfactory locations of all ten acrocentrics were selected for assaying[3]. The probability of blended or unblended satellites in the age groups was determined, controlling for sex, using the formula

$$P = \frac{m}{kn}$$

where *m*—satellite count in the metaphases; *k*—a count of D or G group acrocentrics; *n*—a number of metaphases. Respective relations for groups D and G were

$$P_D = \frac{m}{6n}\ ;\ P_G = \frac{m}{4n}\ .$$

Data were compared using Student's statistics.

Table 5.1.1-1 represents the data on blended and unblended chromatid satellites for the age groups (Fig.5.1.1-2). The analysis showed that men of 48 to 80 years of age had a greater probability of D group chromosome blended satellites, as compared to group G satellites, i.e. *PD* > *PG*. Alternative exceptions were significant. This probability proved insignificant in middle-aged and senile women. The probability of unblended satellites in men and women in all age groups (except for adult men) was greater for group G chromosomes.

Table 5.1.1-2 summarizes the data on the age-related incidence of acrocentric chromosomes with the blended and unblended satellites. In men, D group chromosomes showed a similar probability of having the blended satellites in the control (*c*) and senile (*s*) groups, i.e. *PD* (*c*) = *PD* (*s*). On the contrary, the blended satellites in G group chromosomes were much less likely to occur in the control group: *PG*(*s*) > *PG*(*c*). In women, probability of the blended satellites was similar for group D and G and was much greater at 80 years and older: *PD*(*c*) < *PD*(*s*) and *PG*(*c*) < *PG*(*s*). The chromosome D and G unblended satellites in men and women showed the relation *P*(*c*) > *P*(*s*).

Therefore, the incidence of the blended and unblended chromatid satellites is sex and age dependent.

[3] These selection criteria and culture conditions were employed in all acrocentric chromosome studies described below.

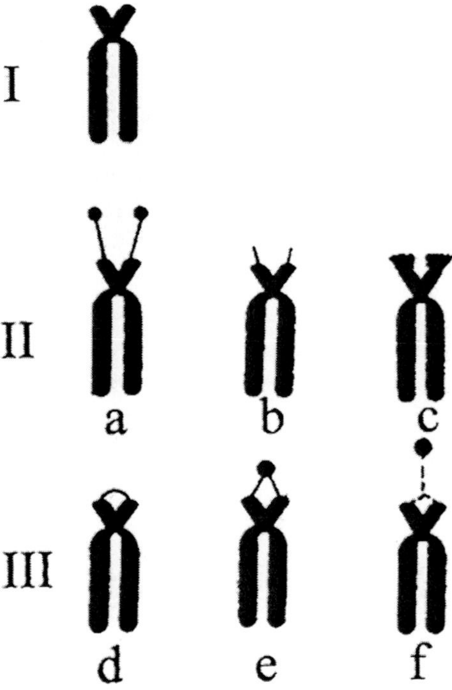

Fig. 5.1.1-1. Types of acrocentric chromosome satellites classification into groups I, II, III: a - classical satellite; b - antenna satellite; c - puff satellite; d - ring satellite; e - blended satellite; f - spheroid satellite.

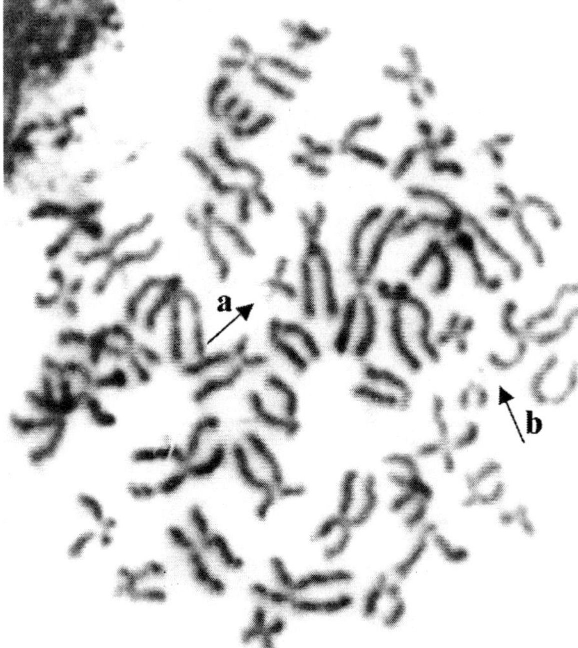

Fig. 5.1.1-2. Metaphase of a 90 year-old woman. Arrows point to (a) group G chromosome with unblended chromatid satellites; (b) group G chromosome with blended chromatid satellites.

Table 5.1.1-1. Probabilities of chromosomes with blended chromatid satellites: comparison within age groups

Age (yr)	No. of Subjects		No. of Metaphases		$P_D = \frac{m_D}{6n}$			$P_C = \frac{m_G}{4n}$			$\frac{P_D - P_G}{\sqrt{\frac{P_D(1-P_D)}{6n} + \frac{P_G(1-P_G)}{4n}}}$			Error level at testing hypothesis $P_D > P_G$		
	M	F	M	F	M	F	Total No.	M	F	Total No.	M	F	Total No.	M	F	Total No.
20-48	6	9	128	136	0,01 (0,13)	0,01 (0,114)	0,01 (0,12)	0,0… (0,14)	0,006 (0,14)	0,003 (0,14)	3,5 (-0,73)	0,42 (-1,57)	1,74 (-1,63)	0,0… (50%)	50% (50%)	50% (50%)
80-114	13	8	419	239	0,014 (0,06)	0,017 (0,08)	0,02 (0,07)	0,01 (0,08)	0,02 (0,13)	0,01 (0,1)	1,82 (-2,3)	0,13 (-4,08)	1,05 (-4,4)	22% (10%)	0,0… (0,0…)	50% (0,0…)

Note. Values in parentheses describe unblended chromatid satellites.

Table 5.1.1-2. Probabilities of chromosomes with blended chromatid satellites: comparison between age groups

Age (yr)	$\frac{P_{D(1)} - P_{D(2)}}{\sqrt{\frac{P_{D(1)}[1-P_{D(1)}]}{6n(1)} + \frac{P_{D(2)}[1-P_{D(2)}]}{6n(2)}}}$		Error level at testing hypothesis $P_1 > P_2$		$\frac{P_{G(1)} - P_{G(2)}}{\sqrt{\frac{P_{G(1)}[1-P_{G(1)}]}{4n(1)} + \frac{P_{G(2)}[1-P_{G(2)}]}{4n(2)}}}$		Error level at testing hypothesis $P_1 > P_2$	
	males	females	males	females	males	females	males	females
80-114 (1)	1,54	2,22	50%	14%	3,5	2,14	0,0…	15%
20-48 (2)	(-4,87)	(-2,89)	(0,0…)	(10%)	(-3,52)	(-2,51)	(0,0…)	(10%)

Note. Values in parentheses refer to unblended chromatid satellites.

5.2. Human Acrocentric Chromosome Associations

5.2.1. Relationship between the Incidence of Recognizable Satellites, Association and Age in Group D and G Acrocentric Chromosomes

The association phenomenon is a highly specific indicator of the nucleolar structure and function at the previous interphase (Hoehn et al., 1971; Hens et al., 1980). It may induce acrocentric nondisjunction during the meiosis or early zygote division, and chromosome rearrangements (Ohno et al., 1961; Ferguson-Smith, 1964). Chromosomes can associate when two chromatid satellites are available (Engman, 1972; Chemitiganti et al., 1984), and so they are defined as associated, when their satellites make up a pair. Therefore, prematurely condensed silver-stained acrocentrics have similar rates of interphase and metaphase associations (Schmiady et al., 1979). In humans, acrocentric associations may be formed by two or more chromosomes (Ferguson-Smith and Handmaker, 1961; Prokofyeva-Belgovskaya, 1986).

Analysis of peripheral lymphocytes showed that ten acrocentric chromosomes may produce 1 to 5 independent associations. Most of cultured lymphocytes carry one association; the incidence of those with two associations is threefold lower. Occasional cells have three, four or more associations. Embryonic fibroblast cultures display a sixfold lower incidence of cells with two associations and no cells with three associations (Prokofyeva-Belgovskaya, 1986). Of ten human satellited chromosomes, both homologous and nonhomologous acrocentrics are involved in association formation (Bishun, 1966).

Association incidence in normal individuals was reported to vary in chromosomes 13, 14, 15, 21 and 22 (Cooke, 1971; Cooke and Curtis, 1974; Schmid and Krone, 1974). Satellite associations in G group chromosomes are more common than in the D group (Lezhava et al., 1972), although autoradiography of D group chromosomes identified the highest association incidence in chromosome 13 (Cooke, 1971).

The association frequency varies in different cells. It is 0.49 per cultured skin cell (Froland and Mikkelsen, 1964), 0.35 in embryonic fibroblasts (Prokofyeva-Belgovskaya, 1986) and 1.32 to 1.46 in lymphocytes cultured from middle-aged humans (Froland and Mikkelsen, 1964; Prokofyeva-Belgovskaya, 1986). The ability of acrocentrics to form associations may also vary within a cell type according to age. It is 46.3 percent in newborns, 87.9 percent at 20 to 40 years and 53.3 percent at 53 to 92 years (Prokofyeva-Belgovskaya, 1986). Similar age-related patterns were reported by other investigators (Aragatsuni, 1967; Lezhava et al., 1972; Mattevi and Salzano, 1975b). More recent evidence has indicated that the incidence of Ag-positive and association-making acrocentrics decreases with age (Liem et al., 1977; Buys et al., 1979; Hansson, 1979; Denton et al., 1981). The incidence of Ag-positive nucleolar organizer regions (NOR) has been correlated with that of satellite associations (Lernid et al., 1980). Changes of acrocentric nuclei or organizers, or else changes of short arm heterochromatin were proposed as explanations for the association formation (Prokofyeva-Belgovskaya, 1986). It was shown by Orey (1974) and Verma and Rodriguez (1985) that the likelihood of acrocentric associations is related to an extent of satellite stalk heterochromatinization. Studies of murine-human somatic cell hybrids suggested that silver staining of short arm acrocentric stalks may provide an indicator of acrocentric chromosome function (Miller et al., 1976; Tantravahi et al., 1977).

The amounts of differential Ag-staining of NOR were found to correlate with the incidence of satellite associations in human acrocentric chromosomes (Miller et al., 1977). Individual counts of stained regions per cell were consistent (5 to 9). The associated chromosomes were usually linked by Ag-stained stalks of variable, sometimes great lengths. Larger areas of staining correlated with a greater incidence of satellite associations.

Evaluation of the association incidence and selective Ag-staining of acrocentric stalks in 5 normal individuals has demonstrated that acrocentric associations selectively involve chromosomes with morphological markers, and short arm stalks in the associations are always Ag-positive (Capoa et al., 1978). Furthermore, D and G group acrocentrics with elongated short arms were reported to have very low incidence of associations; conversely, group Ds+ and Gs+ chromosomes with prominent satellites displayed a strong tendency to make associations ($P < 0.05$ for group D and $P < 0.1$ for group G) (Zankl and Zang, 1974; Phillips, 1975). Elongated satellites on a marker chromosome were documented in a mentally retarded child, with abundant associations made by that chromosome (Vormittag et al., 1972). Similar results were obtained in three families (Capoa et al., 1973); two apparently normal probands examined in each family showed altered chromosome satellites. Although the alterations occurred as structural abnormalities in different D group chromosomes (giant satellites on chromosome 15 in family 1, double satellites on chromosomes 14 and 15 in families 2 and 3), an increased association incidence was seen in all marker chromosomes. The incidence of associations in acrocentric chromosomes with double satellites is much higher compared to acrocentrics without the satellites (Gigliani et al., 1972). In addition, a Q-banding study showed that an acrocentric homolog with a long satellite stalk more commonly formed associations than the other homolog (Schmid et al., 1974).

Acrocentric association formation was found to have the following pattern: 13>21>14>22>15 (Patil and Lubs, 1971). Exposure of the acrocentric chromosomes to trypsin has revealed that chromosome 13 was most commonly and chromosome 15 was occasionally involved in associations (Ardito et al., 1978). Other chromosomes had intermediate rates of the involvement: 13 > 21 > 22 >14 > 15. Homologous and nonhomologous chromosome associations have been examined using differential staining and computer techniques (Galperin-Lamaitre et al., 1977). Homologous chromosomes tended to associate as follows: 13-13 = 21-21 > 14-14 > 22-22 > 15-15 and nonhomologous ones 13-14 > 13 > 15 > 14 > 15; 13-21 > 14-21 > 13-22 > 15-22. Evidence for the nonrandom associations was explained by the molecular organization, structure and function of short arm compartments. It was suggested that ribosomal DNA contents vary within the five acrocentric pairs. Acrocentrics with higher ribosomal DNA levels may associate at higher rates, and these associations show a higher metaphase occurrence.

A family of human repetitive DNA sequences without homologies to Alu, Kpn and alphoid sequences has been identified in humans (Graham et al., 1984). Members of this family are tandem repeats located on chromosomes 1, 13, 14, 15, 21 and 22. These nucleotide DNA sequences are thought to take part in acrocentric satellite association.

There are two interpretations of the associative behaviour of the short arm of a chromosome with or without satellite stalks. A deletion hypothesis suggested a relationship between satellite stalk sizes and numbers of ribosomal cistrones in the nucleolar organizer. Partial deletions and duplications spontaneously occur at a relatively high rate in *Xenopus laevis, Siredon Mexicanum,* and *Bufo marinus* (Schmid et al., 1974).

According to an inactivation hypothesis, nucleolar organizer length is not constant but rather varies with ribosomal cistrone activities in selected nucleolar organizers. Nucleolar organizer inactivation was demonstrated in plant, animal and human cells (Orey, 1974; Schmid et al., 1974; Dumont et al., 1989).

A higher acrocentric association incidence was documented in three generations of subjects with mosaicism of translocated D/G and D/D chromosomes, partial trisomy 21 and X monosomy (Zellweger and Abbo, 1965). It was inferred from these findings that an autosomal dominant gene must be responsible for the higher association incidence in persons with chromosome anomalies.

In a study of Abbo et al. (1966), acrocentric associations were examined in 6 family members, of whom 4 had mosaic cells with several chromosomal anomalies. Chromosome assays in a lymphocyte culture revealed 7 different clones in a 10-year-old girl with Down's syndrome; one of the clones had a normal chromosome set and another showed sex chromosomes of a 45,X type. Autosomal mosaicism 45/46 with a balanced D/D translocation was found in her father and grandmother. A mean incidence of cells with associations was 13.3 percent in control subjects and 26.8 percent in the studied family. It was presumed that a mutant gene must exist in autosomes, one dose of which increases the incidence of acrocentric chromosome associations and a double dose promotes it to a degree causing mitotic abnormalities, with mosaicism as the overall result.

Studies of acrocentric chromosomes (Lezhava, 1979; Lezhava, 1984a) with visible associated or unassociated satellites have focused on the following variables: (1) age-related incidence of acrocentrics with visible satellites; (2) probabilities of association formation in acrocentrics with or without the visible satellites in control and study groups; (3) incidence of D and G group associations in the two age groups.

These variables were examined in two age groups. The criterion was the condition that acrocentric chromosomes were in contact with their satellite stalks.

The D and G group acrocentric chromosomes with and without visible satellites were also evaluated for associations. The probability of the visible satellites in satellite-bearing chromosomes in different ages was determined for

$$P_{(s)} = \frac{m}{n}$$

For D and G group chromosomes this variable was

$$P_{D(s)} = \frac{m_D}{6n}\,;\; P_{C(s)} = \frac{m_G}{4n}$$

where *m*—metaphase amount of visible satellites; *n*—total number of the metaphases. Inter- and intragroup data were compared using Student's statistics. To elucidate the incidence of acrocentrics with visible association-involved satellites, the probability *P*(*A*/*s*)*P*(*A*) (i.e. likelihood of metaphases to have more associations between visible satellites or to be equal to that of all acrocentrics) was examined in both age groups. The *P*(*A*) and *PD*(*A*/*s*) values are asymptotically normal and independent of mean *PD*(*A*), *PG*(*A*) and *PD*(*A*/*s*), *PG*(*A*/*s*) values and dispersion, and these were obtained experimentally. These values will be described separately as

$$P_D(As) = \frac{\sum_{i=1}^{N} k(D)}{6N}; \quad P_g(A) = \frac{\sum_{i=1}^{N} k(G)}{4N}$$

$$P_D(A/s) = \frac{L_s(D)}{m(D)}; \quad P_G(A/s) = \frac{L(G)}{m(G)}$$

where *N*—a total of assayed metaphases; *k*—a number of acrocentric chromosomes involved in associations; *L*—a number of acrocentric chromosomes with visible associated satellites. The findings were compared as follows: *S2P*(*A*)—empirical mean square deviation for the probabilities of association formation by all acrocentrics. For D group chromosomes it was

$$S^2 P_D(A) = \frac{\sum_{i=1}^{N} k^2(P)}{6^2 N} - \left[P_D(A)\right]^2,$$

for G group chromosomes it was

$$S^2 P_G(A) = \frac{\sum_{i=1}^{N} k^2(G)}{4^2 N} - \left[P_G(A)\right]^2,$$

Empirical covariation of estimates:

$$Cov[P^D(A), P^G(A)] = \sum \frac{k(D)k(G)}{6 \cdot 4 \cdot N}$$

Hypothetically, there is no variability of compared data, and so the statistics

$$\frac{P(A/s) - P(A)}{\sqrt{S^2 P(A) + S^2 P(A/s)}} \; \frac{P_G(A) - P_D(A)}{S^2 P_G(A) + S^2 P_D(A) + C_{OV} P_D(A), P_G(A)}$$

must be within a normal standard range.

Table 5.2.1-1 presents the data on visible D and G group acrocentric satellites in both age groups. The analysis suggests that the probability of visible G satellites in individuals aged 80 to 103 years is much higher, as compared to D satellites: *PG*(*s*) > *PD*(*s*). Alternatives were ruled out with a high significance. These probabilities were insignificant in 20- to 48-year-old subjects:

$PD(s) = PG(s)$

Comparative probabilities of acrocentrics with visible satellites in the two age groups are given in Table 5.2.1-2. They proved almost similar for D group chromosomes in the age ranges of 20 to 48 and 80 to 103 years (error level, 16 percent). By contrast, the analysis of the probabilities for G group chromosomes revealed that the hypothesis $PG(s)$ (c) $PG(s)$ (s) had to be rejected in favor of $PG(s)$ $(c) > PG(s)$ (s), with the error level of 3 percent.

Acrocentrics with visible satellites were presumed to be more likely to make associations, and this probability was checked by comparison of respective variables of the hypothesis $PD(A) < PD(A/s)$ and $PG(A) < PG(A/s)$, alternatives being $PD(A)$ $PD(A/s)$ and $PG(A)$ $PG(A/s)$.

The study results showed that D and G group chromosomes in both age groups had a significantly higher probability of having visible satellite associations, as compared to all acrocentrics: $P(A/s) > P(A)$. Alternatives were ruled out with a high significance (99.2 percent).

The 'activity' of D group chromosome association varied with ages, unlike group G. Probabilities of D and G group chromosome associations appreciably differed in persons aged 20 to 48 years (significance, 98.5 percent). The associative behavior of D and G group chromosomes in the old-age group showed no variability (significance, 75.5 percent).

To assess the probabilities of D and G group chromosome associations, it was assumed that they showed no significant difference between the age groups (error level, 25 percent).

5.2.2. Satellite Associations and Ag-Staining Patterns

Acrocentric chromosomes of lymphocytes can produce a variety of associations with various chromosome groups (Fig.5.2.2-1). Associations may be formed by various numbers of acrocentrics (2 to 7) (Fig.5.2.2-2).

Numbers of Ag-positive NOR and associations have been evaluated in chromosomes of 20 lymphocyte cultures obtained by standard techniques from 20 subjects of both sexes ranging in age from 20 to 93 years (Lezhava, 1984*a*). The analysis involved 20 metaphases from 10 subjects (2 women, 8 men) of 80 to 93 years and 10 subjects (3 women, 7 men) of 20 to 50 years. Chromosome preparations were obtained with an air-drying technique. Silver staining was performed as recommended by Bloom and Goodpasture (1976). The G-banding technique of Sumner et al. (1971) was used in chromosome identification. The Ag-bands were visually assessed using a scoring system which described a lacking band as 0, a small band (smaller than chromatid width) as 1 and a large band (chromatid-wide or larger) as 2.

The probability of argentophilic NOR and acrocentric associations (the associations were defined as intersatellite bonds identifiable by Ag-staining) was tested by comparison of two binomials: N and M, which were respectively metaphase numbers in senile and middle-aged individuals. Argentophilic chromosome counts and association counts were denoted n and m, respectively.

Table 5.2.1-1. Testing of the hypothesis that acrocentric chromosomes with visible satellites are more likely to make associations than all acrocentrics: $P(A) < P(A/s)$

Age (yr)	$P_D(A)$	P_D (A/s)	S^2P_D (A)	S^2P_D (A/s)	$\frac{P_D(A/s)-P_D(A)}{\sqrt{S^2P_D(A)+S^2P_D(A/s)}}$	Error level at testing hypothesis $P_D(A) < P_D(A/s)$	P_G (A)	P_G (A/s)	S^2P_G (A)	S^2P_G (A/s)	$\frac{P_G(A/s)-P_G(A)}{\sqrt{S^2P_G(A)+S^2P_G(A/s)}}$	Error level at testing hypothesis $P_G(A) < P_G$ (A/s)
20-48	0.253	0.546	0.0001	0.0038	4.7	0.01%	0.565	0.588	0.0001	0.0087	3.01	0.1%
80-103	0.248	6.630	0.0001	0.0075	4.2	0.01%	0.244	0.660	0.0002	0.0058	5.0	0.01%

Table 5.2.1-2. Probabilities of chromosomes with visible satellites: comparison between age groups

Age (yr)	$P_G(A)$-$P_D(A)$	$COV\ P_G(A), P_D(A)$	$\frac{P_G(A)-P_D(A)}{\sqrt{S^2P_G(A)+S^2P_D(A)+COVP_G(A)P_D(A)}}$	Error level at testing hypothesis $P_G(A)$-$P_D(A)$
20-48	0.02	0.000018	2.94	0.2%
80-103	0.01	0.000013	0.69	accepted

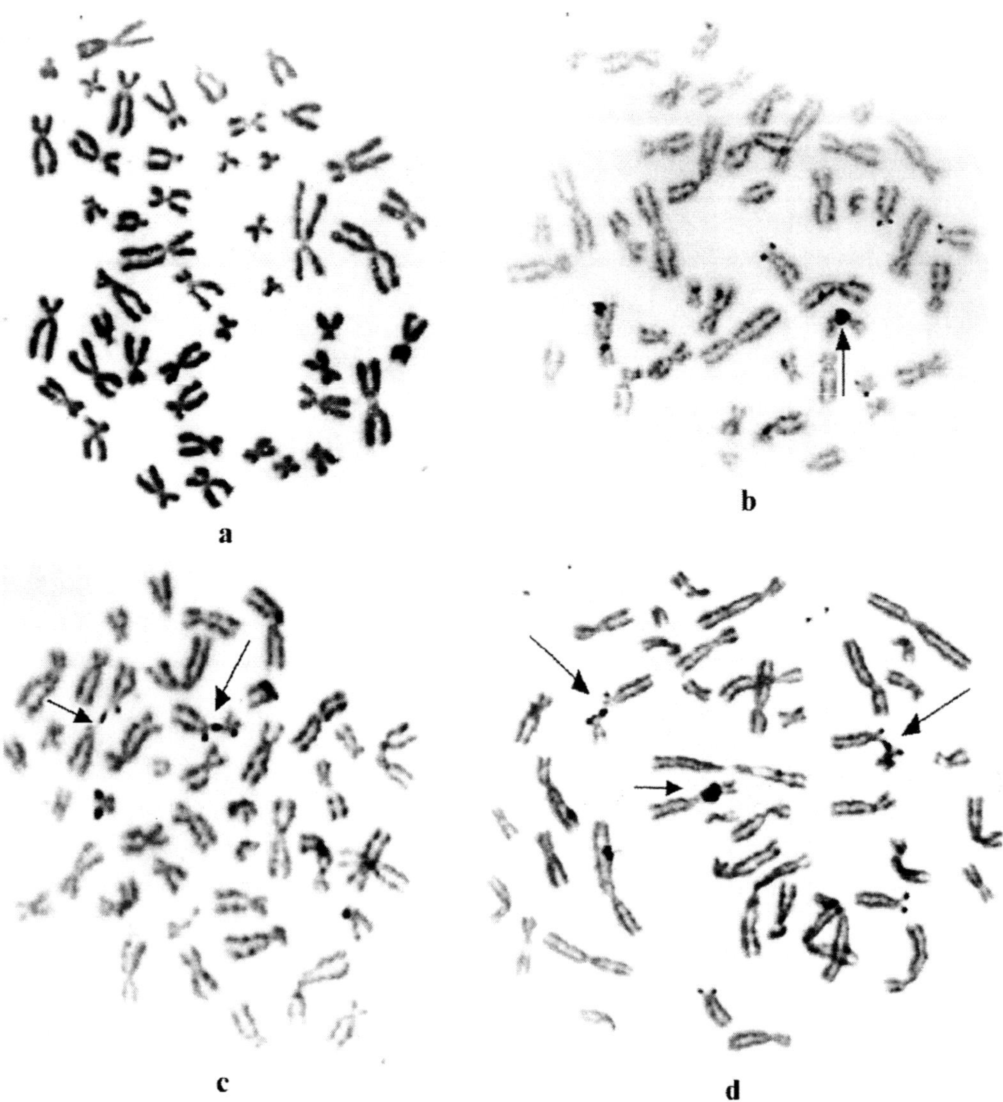

Fig.5.2.2-1. Metaphases of cultured lymphocytes from 80-year-old and older individuals with variable numbers of acrocentric chromosome associations and chromosomes with specific types of associations: a - metaphase without associations; b - metaphase with one association (GG); c - metaphase with two associations (DD, DG); d - metaphase with three associations (DG, DG, DG).

According to the equal probability hypothesis, n should have a hypergeometric distribution with parameters N, M, $n + m$. When $N = M$ and N is a high value, probabilities follow a binomial pattern (1:2), with a $n + m$ number of observations. Normal approximation is shown to have a normal standard distribution (Bolshev and Smirnov, 1968).

Comparative analysis showed that the amounts of silver staining of selected acrocentrics varied according to age. The counts of Ag-positive NOR of all chromosomes (association-involved and nonassociation-involved) in senile age proved significantly lower than those in 20- to 50-year-old individuals (error, 0.99). Moreover, with all interindividual variability, the subjects had consistent NOR numbers per cell (5 to 10).

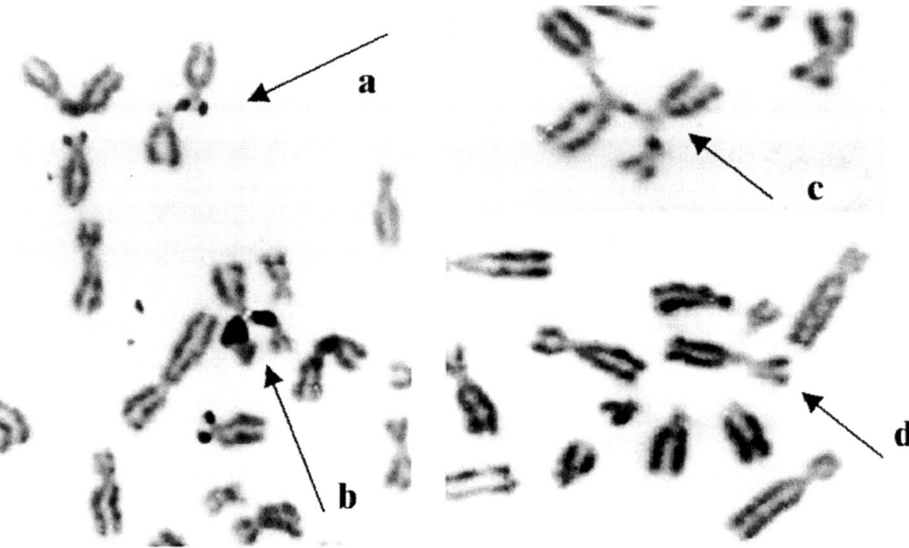

Fig. 5.2.2-2. Chromosome with specific types of associations (peripheral lymphocyte culture): a - two-chromosome, b - three-chromosome, c - four-chromosome, d - seven-chromosome associations.

The equal probability hypothesis was accepted as plausible for argentophilic chromosomes 13, 14, 21 and 22 that were not involved in associations in senile subjects. As the findings were compared to chromosome 15, the hypothesis was rejected in favor of the relation 15 > 14 = 21 = 22 (worst-case error, 4.8 percent).

No significant variability of counts for nonassociation-involved argentophilic chromosomes 14, 15 and 21 occurred in middle adulthood. The probability of nonassociation of chromosome 22 and comparison of it to other nonassociation-involved Ag-positive acrocentrics was assessed; it was found that the equal probability hypothesis had to be rejected in favor of 22 < 14, 22 < 15 and 22 = 21 (greatest error, 1.46).

The hypothesis 13 = 21 must be ruled out for chromosome 13 in favor of 13 > 21 (error, 1.79 percent), while the hypotheses 13 = 14 and 13 = 15 were accepted. The counts of Ag-positive NOR on chromosomes 13 and 21 in senile subjects were decreased relative to the middle-age range. The evaluation of stained NOR sizes using the scoring system (0-1-2) revealed a positive correlation between staining resolutions of association-involved and nonassociation-involved acrocentrics. It is Noteworthy that acrocentrics without staining bands made no associations. By contrast, occurred when chromosomes had even small stained NOR (scoring 1); the frequency of satellite associations increased with staining scores. The frequency of chromosomes with NOR sizes scoring 2 was higher in middle age, as compared to old age.

It is obvious from Table 5.2.2-1 that acrocentric associative behavior is individualized. The activity of association formation by chromosomes 13 and 22 in old age is significantly reduced, compared to middle age. Furthermore, frequencies of satellite associations on homologous and nonhomologous chromosomes suggest a low likelihood of chromosome 13 satellites to associate with chromosomes 21, 22 and 14 in very old ages.

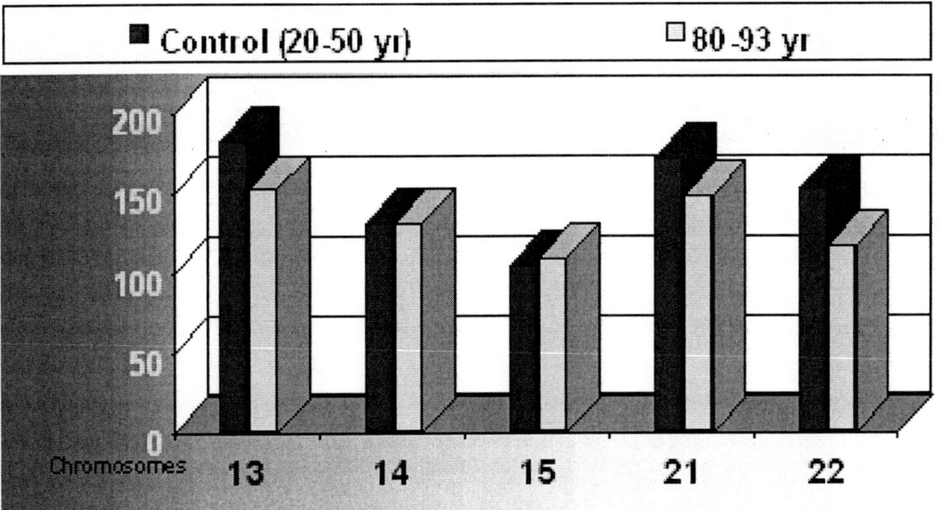

Fig. 5.2.2-3 Alteration of acrocentric associative behavior is individualized with age. 200 metaphases were analyzed in each age group.

The acrocentric chromosomes were evaluated for 'opened' associations, i.e. chromosome attachment by Ag-positive intersatellite stalks, with satellites remaining free (Fig. 5.2.2-4); the "open" association incidence was 85.5 percent in old and 84.1 percent in middle-aged subjects. The total incidence of 'closed' associations, in which no free satellites occurred, was 15 percent. These associations involved two chromosomes in 13.5 percent and three or more chromosomes in 1.9 percent of middle-aged subjects. Respective findings in old subjects were 11.5 and 2.9 percent.

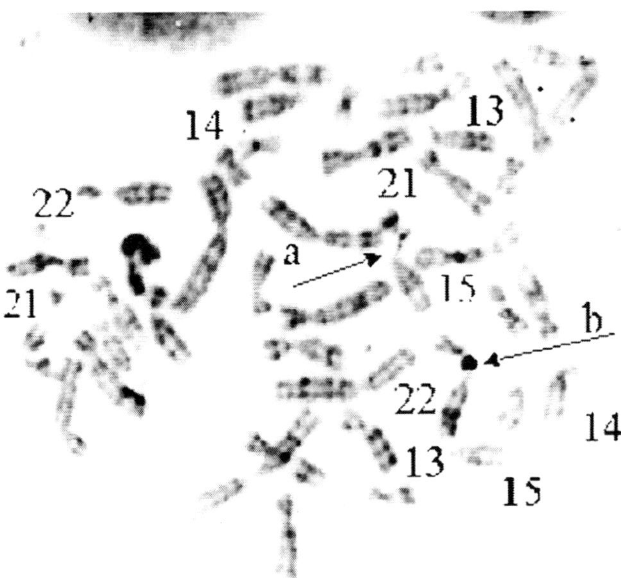

Fig. 5.2.2-4. Association-bearing metaphases with variable sizes of Ag-positive nucleolar organizer regions. Arrows indicate a - "open" association; b - "closed" association.

From these studies we concluded that the intensity of stained NOR and the incidence of acrocentric chromosome associations decline with age, as indicated by significantly lower counts of Ag-positive NOR in acrocentrics 13 and 21 and fewer satellite associations between chromosomes 13 and 21, and 22 and 14. The sizes of staining NOR segments correlated with numbers of nonassociated acrocentrics.

5.2.3. Cis- and Trans-Types of Chromatid Association

Most of acrocentric chromosome associations (85 percent) are formed by single chromatid satellite stalks (Lezhava et al., 1972; Verma, 1983; Lezhava, 1984a). The exposure of lymphocyte cultures to 5-bromodeoxyuridine (BrDU) during two replication cycles revealed two-acrocentric associations that were either at a cis-position (differentially stained acrocentric chromatids with a dark-to-dark or light-to-light association) or a trans-position (chromatids with a dark-to-light or light-to-dark association) (Chemitiganti et al., 1984).

Frequencies of the cis- and trans-orientation of acrocentric chromatid association have been studied by us in very old subjects. Lymphocyte cultures were prepared using a conventional methodology. The study examined 173 metaphases from 9 subjects aged 80 to 89 years and 124 metaphases from 6 subjects aged 20 to 48 years. For differential staining of sister chromatids, BrDU (7.7 μg/ml) was added to the cultures immediately on their initiation.

The lymphocytes were incubated in darkness for 96 h at 37°C. No fluorochromes were used in chromosome staining (Antoshchina and Poryadkova, 1978). Giemsa stain was employed after DNA thymidine was substituted by BrDU. In DNA, thymidine was totally substituted in one of the second-mitosis sister chromatids which stained light and was denoted chromatid 1; only half of DNA thymidine was substituted in the other chromatid which stained dark and was defined as chromatid 2 (Fig. 5.2.3-1). According to association criteria (Lezhava and Khmaladze, 1988b, 1989) the cis-1 position was the term adopted for the light-to-light association, cis-2 position for the dark-to-dark association, and trans-position for the light-to-dark association (Fig. 5.2.3-2).

Table 5.2.3-1 summarizes evidence on cis-1, cis-2 and trans-orientation of chromatid associations in subjects aged 20 to 48 and 80 to 89 years. Statistical analysis of association frequencies proceeded from the assumption that the cis-1 and cis-2 associations have similar chances to occur, and the chances form half of the probability of the trans-oriented association, that is

$$P\text{cis-1}(DD) = P\text{cis-2}(DD) = 1/2\ P\text{trans}(DD) \tag{1}$$

$$P\text{cis-1}(GG) = P\text{cis-2}(GG) = 1/2\ P\text{trans}(GG) \tag{2}$$

$$P\text{cis-1}(DG) = P\text{cis-2}(DG) = 1/2\ P\text{trans}(DG). \tag{3}$$

These equalities represent the hypothesis that chromatids-1 and chromatids-2 participated in the association with the same probability.

The data of the middle-aged group fitted the hypotheses (2) and (3). The statistics

$$X^2(GG) = \frac{(V_{cis-1}(GG) - V(GG)/4)^2}{(1/4)V(GG)} + \frac{(V_{cis-2}(GG)) - V(GG)/4)^2}{(1/4)V(GG)} + \frac{(V_{trans}\,GG) - V(GG/2)^2}{(1/2)V(GG)}$$

An important consideration is deviation of the data from the hypotheses (1)-(3). The deviation suggested that chromatids 1 and 2 of D chromosomes had different associative activities, unlike G-chromosome chromatids. Indeed, if D-chromosome chromatid 2 were more active than chromatid 1, probabilities should be

$P\text{cis-1}(DD) < 1/2\,P\text{trans}(DD) < P\text{cis-2}(DD)$

$P\text{cis-1}(DG) < 1/2\,P\text{trans}(DG) < P\text{cis-2}(DG)$

and these agreed well with the actual findings.

In conclusion, sister chromatids of acrocentric chromosomes show a functional heterogeneity in very old subjects. D-chromosome chromatids 1 with 5-BrDU in all chromosomal DNA participate in associations less actively than chromatids 2 with 5-BrDU incorporated in half of the DNA.

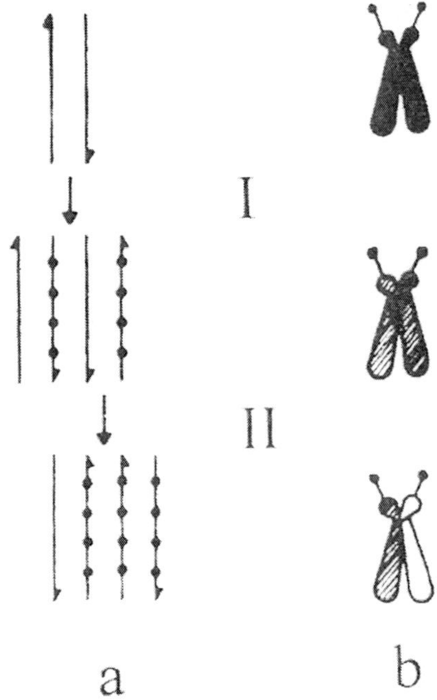

Fig. 5.2.3-1. BrDU incorporation in DNA during two replication cycles: a - DNA molecule strands; b - chromosomes I, II in the 1st and 2nd mitosis in the presence of BrDU. 1 - DNA strands without 5-BrDU; 2 - DNA strands with 5-BrDU; 3 - chromatids without 5-BrDU; 4 - chromatids with 5-BrDU incorporated by half of the DNA strands; 5 - chromatids with 5- BrDU incorporated by all DNA strands. BrDU; 2 - DNA strands with BrDU; 3 - chromatids without 5-BrDU; 4 - chromatids with 5-BrDU incorporation by half of DNA strands; 5 - chromatids with 5- BrDU incorporated by all DNA strands.

Table 5.2.3-1. Chromatid association of acrocentric chromosomes in cis- and trans-orientation

Age (yr)	No. of subjects	No. of cells	No. of association	No. of cells without association	Orientation of association chromatids	Association with single satellite Chromatid stalks			Association with double chromatid satellite stalks			Association with single and double chromatid satellite stalks		
						DD	DG	GG	DD	DG	GG	DD	DG	GG
80-89	9	173	197	37	Cis-1*	12	11	3				14	14	7
					Cis-2**	22	44	14				24	47	8
					Trans***	23	47	14	1	5	2	25	57	18
20-48	6	124	148	16	Cis-1	10	16	3				16	20	6
					Cis-2	13	17	5				19	21	8
					Trans	13	37	9	3	8	1	19	53	11

* In two differentially stained acrocentric chromatids a light-to-light association was registered as the cis-1 position.
** A dark-to-dark association was registered as the cis-2 position.
*** Dark-to-light or light-to-dark alignment was termed as trans-position

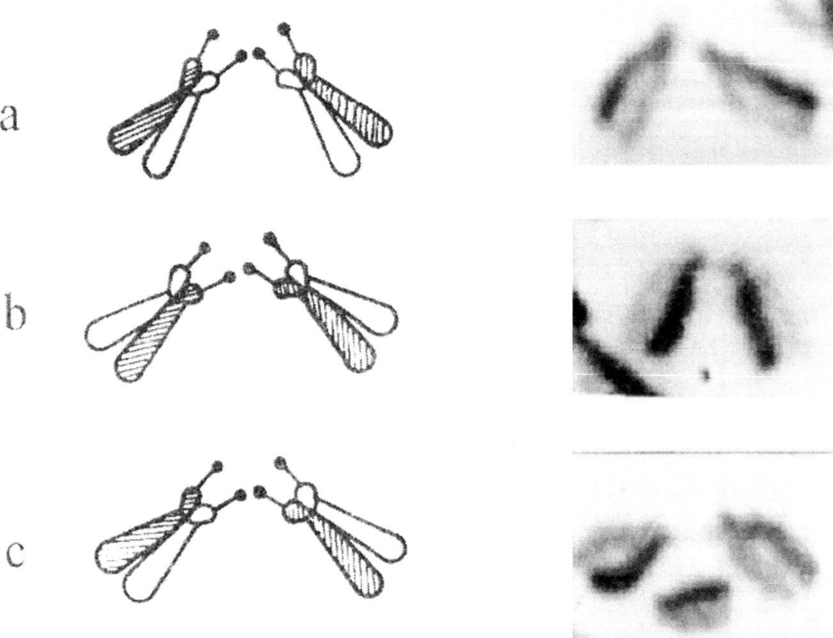

Figure 5.2.3-2. Associations of acrocentric chromatid satellite stalks. a - cis-1 position (light -to-light chromatid association); b - cis -2 position (dark-to dark association); c - transposition (dark-to-light association).

5.3. Transcriptional Activity of Ribosomal Cistrones

Our studies have examined endogenous RNA synthesis during ribosomal gene activation in blast-transformed lymphocytes from 72- to 76-year-old subjects (Lezhava and Dvalishvili, 1992).

Twelve lymphocyte cultures were obtained from 4 donors aged 30 to 45 years and 4 donors aged 72 to 76 years. Two parallel assays were used for each lymphocyte sample. The rates of RNA synthesis in the samples were judged by radioactivity recovery in acid-insoluble material, expressed in impulses per minute per amount of DNA.

The lymphocytes were incubated with four nucleoside triphosphates: adenosine triphosphate (ATP), guanosine triphosphate (GTP), cytosine triphosphate (CTP), and uridine triphosphate (UTP). One of these was labeled with radioactive carbon. After incubation, cold trichloracetic acid (TCA) was poured into the culture; if the nucleotides were bound and RNA was produced, the label was detected in the newly produced polymer in the acid-insoluble sediment. The amount of radioactivity was a measure for RNA synthesis intensity in the sample.

Equal volumes of the culture with suspended lymphocytes were placed into 8–10-ml test tubes for the parallel assays. After 10-min centrifugation at 2400g supernatants were removed, and pellets were submerged in the incubation medium which contained 50 μM of tris-HCl (pH 8.3), 7.5 μM of MgCl2, 0.1 μM portions of UTP, GTP, CTP (Renal, Hungary), and ^{14}C-ATP (0.31 μCi/mM). The lymphocyte samples were incubated for 20 min at 37°C. The reaction was quenched by cooling of the test tubes to 0°C and addition of 5 ml of 5 percent TCA solution containing 0.02 percent of sodium pyrophosphate and 200 μg of an albumin substrate. The samples were stored in cold for at least 40 min, sediments were collected on membrane filters (Hufs, Czechoslovakia, pore size 0.17 μ), washed twice in a funnel with 5 percent TCA with pyrophosphate and once with ethanol; after air-drying radioactivity was measured using a SL-30 scintillation counter (France).

To measure RNA polymerase activity in the cell suspension, DNA content had to be determined in a corresponding volume of the suspension. The DNA content was determined as follows: two samples of the culture mixture were centrifuged at 2400g for 10 min. Pellets were washed with an ethanol-chloroform mixture (3:1) and ether (by sample suspension at room temperature) and then centrifuged. This treatment removed low-polymer acid-soluble agents and lipids from the cells. The pellets were placed in 1 ml of 1N KOH and stored for 1 h at room temperature. Hydrolysate was neutralized by addition of 2 ml of 0.5 N HClO4, stored for 1 h in cold and centrifuged for 10 min at 2400g. Supernatants, which were RNA hydrolysates, were removed, and DNA was measured in pellets, with 3 ml of 0.5N HClO4 added to them. The test tubes were stoppered with condensers and stored in the water-bath for 30 min at 90°C. Upon cooling, the mixture was centrifuged, and the DNA was spectrophotometrically determined in supernatants (DNA hydrolysates) at 270-290 nm. The DNA levels were calculated from gradients of UV absorption.

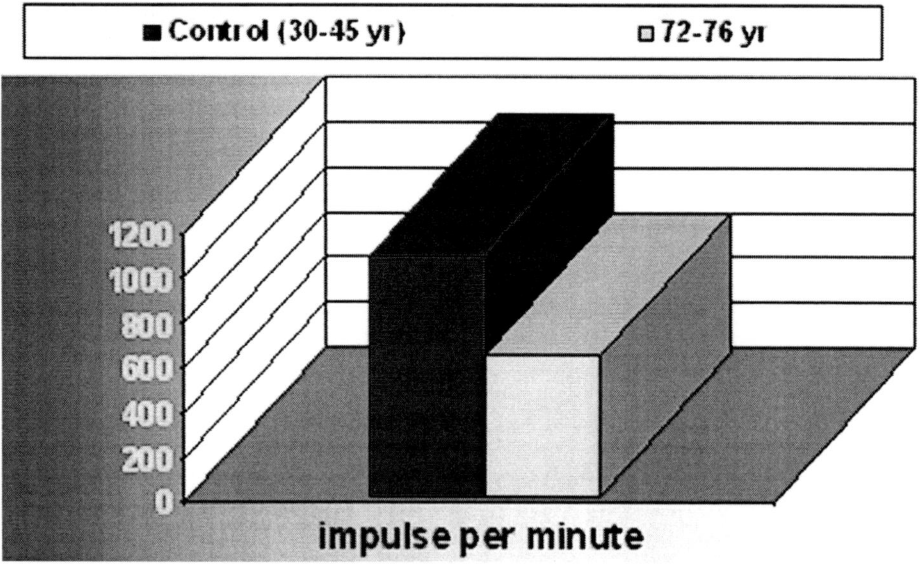

Fig. 5.3-1. Transcriptional activity of blast-transformed lymphocytes labeled with 14C–ATP cpm/min on 15 g DNA in aging

The data of Table 5.3-1 are means of two parallel assays (minus radioactivity of nonincubated samples). The transcriptional activity of ribosomal cistrones in lymphoblasts from subjects aged 30 to 45 and 72 to 76 years indicates variability of DNA-dependent RNA polymerase activity. After 20-min incubation of the samples at 37 °C in the label-containing environment, transcriptional rDNA activity decreased with increasing durations of culture and greater ages of the donors' cells. In lymphocytes of old subjects, the endogenous RNA polymerase activity of ribosomal genes proved lower than in the control group. Radioactivity per 15 μg of DNA in cultured cells of old subjects was from 668 to 721 cpm/min (control, 48-h cultures—from 1020 to 1120 cpm/min). The transcriptional activity in individuals was fairly stable in spite of its intraindividual variations.

DISCUSSION

The results of these studies indicated that the incidence of acrocentric chromosomes with prominent associated or unassociated satellites and the transcriptional activity of ribosomal cistrones are variable in old subjects. Cell system ontogenesis may alter chromosome telomeres, in particular the morphology of acrocentric satellites (Prokofyeva-Belgovskaya, 1986). Our studies suggest that this alteration specifically occurs as prominent D and G chromosome satellites (Lezhava and Khmaladze, 1988b, 1989; Lezhava and Dvalishvili, 1992). Metaphase assessment in human lymphocyte cultures identified a variety of satellite types (puff-like, ring-shaped, blended, spheroid). Some of them were found in plants (Deryagin and Iordansky, 1971) and human tumor (Heliot et al., 2000).

Evidence for a higher incidence of communicating chromatid satellites of group D and G chromosomes in aging women suggests that this phenomenon is a factor in births with Patau's and Down's syndromes, which are chromosomal anomalies associated with chromatid nondisjunction of D and G chromosomes at the second meiotic anaphase or zygote division. The phenotypical lesion of the satellite region is thought to be manifested by its functional variability (Deryagin and Iordansky, 1971; Verma and Rodriguez, 1985).

Short arms, satellite stalks and satellites of human acrocentric chromosomes are heterochromatic regions. The homology of all heterochromatic regions explains their ability to align (Prokofyeva-Belgovskaya, 1986). The alignment of acrocentric chromosome short arms underlies the association formation. In our studies, acrocentrics with the prominent satellites showed a significantly higher probability of association than all acrocentric chromosomes, indicating a primary role of the satellites in D and G chromosome association in all age groups. Similar results have been reported by Engmann (1972) and Dumont et al. (1989).

The associative patterns of D and G group chromosomes varied in two age groups involved by our study. This variability has been observed by others in middle-aged individuals (Cohen and Shaw, 1967; Zang and Back, 1968; Galperin-Lemaitre, 1977; Ardito et al., 1978).

The counts of Ag-positive NOR and frequency of associations produced by stained acrocentric stalks at 80 years and older were significantly lower than in middle age (Lezhava, 1984a). The two variables are known to correlate (Galperin-Lemaitre et al., 1980) and to be genetically determined (Kirsch-Valders et al., 1980; Egolina et al., 1981; Taylor and Martin-De Leon, 1981; Verma and Rodriguez, 1985). Also, they appear to contribute to the nucleolus formation. Since over 80 percent of observed associations are "opened" and multiple nucleolated cells have no associating acrocentrics, the "opened" associations may be presumed to be a functional analogue of a single acrocentric chromosome with Ag-positive NOR (Lezhava, 1984a).

A progressive decrease in the incidence of Ag-positive NOR with advancing age (Buys et al., 1979; Denton et al., 1981) agrees with our findings (Lezhava, 1984a). In our experience, a significant decrease in NOR staining with age involved mostly D and G group chromosomes 13 and 21.

As regards the contribution of satellite-bearing chromosomes to the nucleolus formation, correlations between the frequency of Ag-positive NOR and satellite associations, and the rates of NOR staining observed by us were consistent with reported evidence (Lernid et al., 1980; Galperin-Lemaitre et al., 1980).

Our study has provided another confirmation of the fact that the frequency of associations initially increases in lymphocyte cultures and subsequently decreases with senescence (Prokofyeva-Belgovskaya et al., 1986; Aragatsuni, 1967; Lezhava et al., 1972; Mattevi and Salzano, 1975b; Lezhava, 1979, 1984a; Kadotani et al., 2002). The decrease occurs because senescence is associated with declining ability of D and G group acrocentrics to form associations (Mattevi and Salzano, 1975b; Liem et al., 1997; Lezhava, 1979). The Ag-staining technique allowed us to demonstrate that the likelihood of chromosomes 13 and 22 to enter associations is lower in very senile age.

In addition, we found the transcriptional activity of RNA polymerase I to be depressed in lymphocytes from old subjects: radioactivity per 15 μg of DNA was 668 to 721cpm per minute (versus 1051 to 1120 cpm per minute in controls). Similar impairment of the activity was seen by other investigators. Biochemical and autoradiographic studies identified low RNA synthesis rates in human lymphocytes in vitro aging (Lezhava and Dvalishvili, 1992) and cultured WI-38 fibroblasts (Ryan and Cristofalo, 1975).

The frequency of acrocentric chromosome associations in humans is known to be related to the rates of condensation (heterochromatinization) of satellite stalks (Orey, 1974; Schmid et al., 1974). Acrocentrics with elongated short arms make little associations, while D and G group chromosomes with well-defined elongated satellite stalks strongly tend to be associative (Lezhava et al., 1972; Orey, 1974; Miller et al., 1977). Orey (1974) classified acrocentric chromosomes of the G group into five types: (1) a (G*s*-) type, which is devoid of the stalk, i.e. its satellite stalk is heterochromatinized. In these chromosomes, few, if any, rRNA genes of nucleolar organizer regions are accessible to transcription. In chromosomes of types 2 and 3, deheterochromatinization by satellite stalks increases the number of rRNA genes available for active transcription. The increase in the synthetic activity of rDNA is parallelled by a higher satellite association. Types 4 and 5 (G*s* and G*s*+) are chromosomes with maximal deheterochromatinization of satellite stalk and, respectively, maximal rRNA synthesis. Therefore, the variation of the satellite stalk lengths was interpreted as a function of synthetic activities of ribosomal cistrones (Orey, 1974; Lezhava and Dvalishvili, 1992).

The rDNA tandem arrangement in S.cerevisiae induces transcriptional silencing of RNA polymerase II-transcribed genes. This Sir2-dependent form of repression (rDNA silencing) also functions to limit rDNA recombination and is involved in lifespan control (Buck et al., 2002). Silencing of polymerase II-transcribed genes intergrated within the rDNA silencing is distinct from TPE (telomer position effect) and HM silencing. Few genes other than Sir2 have so far been linked to the rDNA silencing process (Smith et al., 1999).

Subsequent immunolocalization revealed that the Sir4 complex no longer resided at the telomere but was relocalized to the nucleolus. Sir4 essential for other silenced loci paradoxically inhibits rDNA silencing. rDNA silencing strength directly correlates with cellular Sir2 protein (sir silent information regulator) levels. The endogenous level of Sir2p was show to be limiting for r DNA silencing. Furthermore, small changes in Sir2p level altered rDNA silencing strength. Sir4 inhibition of rDNA silencing is mediated through Sir2. Furthermore, r DNA silencing is insensitive to Sir3 overexpression. This negative effect of Sir4 overexpression was overridden be co-overexpression of Sir2 (Smith et al., 1999)

Furthermore, chromosomes with large Ag-bands in NOR associate at higher rates (Miller et al., 1977). There is a positive correlation between the NOR staining intensity and lengths of acrocentric satellite stalks (Lernid et al., 1980; Taylor and Martin-De Leon, 1981; Heliot et al., 2000). Consequently, a long satellite stalk has more active ribosomal genes (Dittes et al., 1974; Goodpasture et al., 1976) and enters associations at a higher rate (Orey, 1974; Miller et al., 1977). The implication is that lower numbers of chromosomes with Ag-positive NOR and a decrease in association frequency and rDNA transcriptional activity in very old age reflect a heterochromatinization-dependent change in satellite stalk lengths.

A noteworthy finding of our study is the differing associative ability of sister chromatids 1 and 2 of cis-positioned D group chromosomes. Chromatids 1 (with 5-BrDU incorporated by all DNA strands) less actively produce associations than chromatids 2 (with 5-BrDU in only half of the strands); this makes up the functional difference between satellite stalks of the sister chromatids. The decrease in the ability of sister chromatid stalks of D group chromosomes to make associations is related to a change in the functional regulation of ribosomal cistrones. The change seems to be induced by phenotypical variability (heterochromatinization) of the stalks in the aging process (Lezhava, 1984a; Lezhava and Khmaladze, 1989; Lezhava and Dvalishvili, 1992; Lezhava, 1996, 2001a, b).

CONCLUSION

Various types of blended and unblended chromatid satellites were identified on D and G group acrocentric chromosomes in individuals of 20 to 80 years or older. The blended satellites were more common in older women. It appears that the chromatid satellites either come in touch with or blend to cause chromatid nondisjunction at the second meiotic anaphase or zygote mitosis.

The probability of association formation by D and G group acrocentric chromosomes varies in very old subjects; satellite stalks have a primary role in association. Moreover, sister chromatids of acrocentric chromosomes displayed functional heterogeneity in very old subjects. The frequency of associations produced by silver-stained acrocentric stalks and numbers of Ag-positive NOR in individuals aged 80 or older were statistically significantly lower than in middle age. The transcriptional activity of ribosomal cistrones also declined with age. The aging-related heterochromatinization of acrocentric satellite stalks appears to account for this.

CHAPTER VI

CHROMOSOME ARRANGEMENT

6.1. CHROMOSOME ARRANGEMENT IN METAPHASE SOMATIC CELLS

The eukaryotic nucleus represents a complex arrangement of heterochromatic and euchromatic domains, each with their specific nuclear functions. Somatic cells of a multicellular organism are genetically identical, yet they may differ completely in nuclear organization and gene expression patterns.

Early studies on the nuclear architecture of somatic cells in different species of plants, *Hemiptera*, amphibians and *Drosophila* indicated that comparably sized homologous chromosomes tend to pair in mitotic metaphases (Stevens, 1908). At that time, the paired arrangement of homologs was thought to be a principal pattern of the nuclear chromosome location (Metz, 1916; Zhivago, 1928). However, the spatial proximity of the homologous chromosomes was challenged by Navashin (1926). His analysis of *Crepis capillaris* 768 metaphases with three chromosome pairs showing much difference indicated that the paired homologous association occurred only in 14.6 percent of metaphases, i.e. most of the chromosome arrangement during the cell division was random. The issue was pursued by Navashin in 1947. Element analysis of 16 cyclical relocations of *Crepis capillaris* chromosomes yielded a probability for each of the relocations to occur during a random combination of elements. Proof of the random arrangement was obtained by Navashin in his study of 9,198 somatic metaphases. The evidence that chromosomes do not have a fixed position in the nuclei, was reported in experiments of in situ hybridization and three-dimensional reconstruction (Manuelidis and Borden, 1988). More recently, fluorescence in situ hybridization (FISH) studies also showed that in human cells there is a relatively random array of chromosomes on the mitotic ring (Allison and Nestor, 1999). However, Zhivago argued, in 1928, that the arrangement of homologous chromosomes is not random but results from a change of the paired pattern to a mirror one in the metaphase. Homologous pairing was seen by Kitani (1963) in *Crepis capillaris* and by Feldman et al. (1966) in wheat rootlet cells. The conclusion of the latter study was that homologous association is prominent at the late anaphase or early telophase, retained during the interphase, and weakened at the metaphase because of a nuclear membrane rupture.

Electron microscopy data from human fibroblast cells also indicated the occurrence of a non-random spatial positioning of the chromosomes (Mosgöller et al. 1991). Fluorescence in situ hybridization (FISH) studies in prometaphase rosettes of human cells also showed that chromosomes are fixed with the homologues separated from each other by at least 90° (Nagele et al. 1995).

It was suggested that to produce the association, adjacent centromeres of homologous chromosomes come in contact with the nuclear membrane, the spindle being a major mediator of the process. Somatic association of homologs occurs as a circular displacement of centromeres at the early telophase (Avivi et al., 1969; Tamayo, 2003a). These results have been obtained on the basis of both topography of chromosomes at nanometer scale in air and in physiologic solution using atomic force microscopy. Imaging of the dehydrated structure in air indicated radial arrangement of chromatin loops as the last level of DNA packing (Tamayo, 2003a).

A mutual chromosome association and its role in the maintenance of the spatial nuclear organization have been examined in sexually mature mice (Monakhova and Steppe, 1975; Monakhova, 1989).In metaphases of dividing spermatogonia, all 40 chromosomes tended to have a circular clustering of centromeres, making a tight centromeric association. Electron microscopic studies of the investigators confirmed that the primarily centromeric regions were communicated into a strong association by a fibril system. It was presumed that the integration of all chromosomes into a system by the centromeric ring is a function served by satellite DNA which is abundant in centromeric heterochromatin of murine chromosomes and able to produce permanent associations.

As Homolog pairing was studied in *Triticum aestivum* by Feldman et al. (1966), the somatic association was found to be controlled by a gene system located on homologous chromosomes 5A, 5B, and 5D. The long arm of chromosome 5B (5Bq) carries a gene (s) which suppresses homologous chromosome association, thus tending to induce a random homolog arrangement in the somatic cell nucleus. Alternatively, the short arm of chromosome 5B (5Bp) and the long and short arms of chromosomes 5A and 5D carry genes for somatic homologous association. The assessment of 5B gene effects on the arrangement of homologs in the mitotic metaphase suggested that they are located closer in the nucleus than homologs. Relative locations of membrane sites of contact with homologs are altered by genetic factors. It was demonstrated by Singh et al. (1976) that somatic association of homologous chromosomes in the interphase nucleus is a requisite for chromosome pairing during meiosis. The chromosome rearrangements lead to aneuploidies in mitosis and meiosis. It was suggested that somatic pairing, endomitosis, meiotic alterations, and chromosomal aberrations can be correlated processes (Becak et al., 2003).

It was also found that the location of a chromosome is related to its size. It was recognized that larger chromosomes usually reside in the metaphase periphery and small ones cluster in the metaphase center. Similar results were obtained in plant, avian, reptilian, mammalian and human cells (Zhivago, 1928, Evans and Swery, 1929; Andres, 1934; Sokolov, 1937; Sosa and Cories, 1984; Hilliker and Appels, 1989; Greaves et al., 2003).

The peripheral metaphase location of large chromosomes and the central one of small chromosomes were documented by a fluorescent Q-banding technique (Warburton et al, 1973). Lymphocytes from 21 donors were cultured for 48h and then fixed using a standard technique without hypotonic treatment and colchicinization. A study using G banding of chromosomes in lymphocyte cultures showed that chromosome size and heterochromatin content influence the chromosome location in the metaphase, while the centromere's position does not (Ford and Lester, 1982). Peripheral locations were delineated for chromosomes 1, 3, 16 and the X chromosome (Hoehen and Martin, 1973; Miller et al., 1963a, b; Barton et al., 1965; Kowarzyk et al., 1967; Warburton et al., 1973; Tamayo, 2003b).

Furey and Haussler (2003) developed a dynamic programming algorithm that employs results from approximately 950 fluorescence in situ hybridization experiments to approximate the locations of the 850 high-resolution bands in the June 2002 version of the draft human genome sequence. These band predictions support previously identified correlations between band stain intensity and certain structural characteristics of chromosomes, namely GC content, repeat structure content, CpG island density, gene density and degree of condensation.

The actual spatial chromosome arrangement in the cell nucleus may be correlated with the usual planar pattern seen in the metaphases (Budyakov and Zolotarev, 1971; Sosa and Cories, 1984; Joachimiak, 1987; Hiraoka et al. 1990; Heliot et al., 2000). It has been shown that in the interphase, the relative positions of the chromosomes may be affected by gene activity (Manuelidis and Borden, 1988).

Homologous pairing was described in metaphase fibroblasts of embryonic *Muntjacus muntjak* cultures (Heneen and Nichols, 1972).

An association of nonhomologous chromosomes may occur because of genetic homology of their heterochromatic regions, resulting in structural mutations (Grell and Day, 1970; Chadov and Chadov, 1977, Sele et al., 1977; Kubai, 1987; Schmid et al., 1989). Genotypes with a tendency to mitotic pairing of nonhomologous chromosomes were identified in *Drosophila* (Grell, 1967). Mitotic nonhomologous pairing induced multiple spontaneous exchanges in pre-meiotic sex cells of *Drosophila* (Chadov, 1975).

This spectrum of evidence suggests that cells of different origins have at least three types of nuclear somatic association of chromosomes and their elements:

(1) homologous chromosome association;
(2) nonhomologous chromosome association;
(3) association of all chromosomes' centromeres in the polar ring. These associations make a basis for spatial chromatin arrangement in the nucleus (Svidchenko, 1975).

The literature concerning homologous pairs in humans is rather controversial. Intercentromere distances were evaluated in lymphocyte and skin cultures of normal men and women; the proximity of well-identifiable homologs of chromosomes 1, 2, 3 and 16 was first recognized in this study (Schneiderman and Smith, 1962). Homologous pairs of all 23 chromosomes were found in 70 metaphases from females (Barton et al., 1963). However, no pairing was seen in males and in metaphases with abnormal chromosome numbers unrelated to sex. Homologous pairing between chromosomes 1, 2 and 3 was confirmed in a study of 62 metaphases (Merrington and Penrose, 1964). Other investigators reported no significant variability of distances between homologous and nonhomologous chromosomes (Kowarzyk et al., 1967). A lack of homologous pairs was reported by the study of Sele et al. (1977) which employed G-banding to identify all chromosomes in 400 metaphases of lymphocyte cultures. However, Hens et al. (1975) reported that chromosomes 1, 17, 19 and 20 and acrocentric chromosomes did tend to make homologous pairs in 50 trypsin-treated, G-banded metaphases from normal 20-year-old women.

According to Sun et al. (2000) two biophysical models could explain chromosome positioning. In the volume exclusion model, non-random size dependent spatial positioning is explained by steric hindrance among chromosomes. In the mitotic preset model confirmed by their findings, the size-dependent positioning is established and maintained in mitosis and G1 interphase. In addition, mutagen-induced variability of chromosome locations in lymphocytes was documented in vivo (Kirsch-Volders et al., 1978, 1980; Verschaeve et al., 1978).

6.1.1. Distribution of Homologous Chromosome Pairs 1, 2, 3 and 16

The spatial relationship of homologous chromosome pairs 1, 2, 3 and 16 has been statistically evaluated using a mathematical model. The in vitro study examined 384 metaphases of lymphocytes from 40 subjects, of whom 20 were apparently healthy ones from 80 to 114 years of age, and 20 were control subjects from 20 to 48 years of age. The numbers of men and women and numbers of metaphases donated by them were equal in both groups.

Each metaphase was examined for an ordered cyclical arrangement of assayed chromosomes 1, 2, 3 and 16 that were numbered in such a way that chromosomes 1, 2, 3, 4, 5, 6, 7 and 8 made up groups of homologous pairs. To test the hypothetical nonrandom mutual positions of homologous chromosomes, neighboring homologous chromosomes were counted in each metaphase. The test compared the empirical distribution of these numbers with the theoretical distribution deduced with an assumption of a random chromosome arrangement. The comparison relied on formulae of Chitashvili (Lezhava and Chitashvili, 1979). Distances between the centromeres of each pair (i.e. a number of chromosomes occurring between the homologs) were measured in order to explore a relationship between proximity of the homologous pairs and chromosome types.

These tests reliably elucidated consistent patterns of mutual homolog positions in the senile and control subjects. Tables 6.1.1-1 and 6.1.1-2 present the results of this study. Table 6.1.1-1 shows that the data of a cumulative control sample significantly deviate from the random distribution hypothesis (RDH) in favor of homologous chromosome proximity (Fig. 6.1.1-1); with three degrees of freedom, $X2 = 25.0$ in the sample corresponded to a 99 percent significance. A similar deviation from RDH was seen in women ($X2 = 12.5$; significance, 99 percent) and men ($X2 = 11.8$).

By contrast, the data of old subjects did not show a significant deviation from RDH (Table 6.1.1-1), i.e., homologous chromosomes were not likely to be close in this group.

Table 6.1.1-2 shows *Li* distances between homologous chromosomes of each of the four pairs and expected RDH values of these.

Comparison demonstrated that the significant deviation from RDH in the cumulative control sample involved homologous chromosome pairs 1 (1 and 2) and 16 (7 and 8), with $X2$ respectively 11.4 and 15.0 (significance, 99 percent).

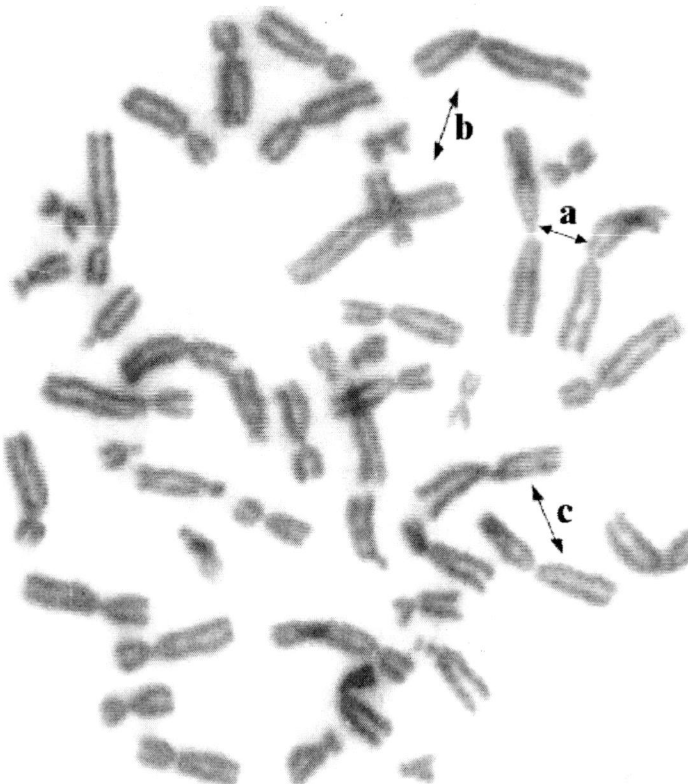

Fig. 6.1.1-1. Metaphase of a 25-year-old man. Arrows show: a - homologous chromosomes A1; b - homologous chromosomes A2; c - homologous chromosomes A3.

No significant deviation from RDH was seen in pairs 2 (3 and 4) and 3 (5 and 6). In women of the control group, significant proximity was found only in homologous pair 16 (chromosomes 7 and 8) with $X2 = 9.1$ (three degrees of freedom, significance, 98 percent); in control group men, proximity occurred in pairs 1 (chromosomes 1 and 2) and 16 (7 and 8), with $X2$ 8.6 and 8.1 respectively.

In the old group, lymphocyte metaphases displayed a significant deviation from RDH toward proximity of homologous pair 2 (chromosomes 3 and 4) in women ($X2 = 7.6$; significance, 94 percent) and pair 16 in men ($X2 = 7.0$). Other homologous pairs showed no significant deviation from RDH, suggesting that senile age is associated with the most efficient homolog separation in pairs 1 and 3 in persons of both sexes, pair 16 in women and pair 2 in men.

To sum up, median distances between homologous chromosomes were appreciably increased (except for pair 2 in women and pair 16 in men) in metaphases of lymphocytes from individuals aged 80 to 114 years.

Table 6.1.1-1. Numbers of identified homologous chromosomes

Age (yr)	Sex	Homologous chromosome pairs					x^2
		n^0	n^1	n^2	n^3	n^4	
	F	26(29)	35(36)	25(22)	8(7)	2(2)	0,8
80-110	M	30(29)	43(36)	13(22)	10(7)	0(2)	5,2
	F, M	56(58)	78(72)	38(44)	18(14)	2(4)	2,6
	F	19(29)	37(36)	22(22)	12(7)	6(2)	12,5
20-48	M	24(29)	27(36)	28(22)	12(7)	5(2)	11,8
	F, M	43(58)	64(72)	50(44)	24(14)	11(4)	25,0

DISCUSSION

There is a great deal of evidence to confirm an ordered arrangement of chromatin in interphase nuclei (Sosa and Cories, 1984; Joachimiak, 1987; Schmid et al., 1989; Mosgoller et al., 1991; Sun et al., 2000; Fransz et al., 2003). Comings (1968) proposed that four major determinants of it are (1) association of chromatin with the nuclear membrane; (2) nonrandom distribution of metaphase chromosomes; (3) nonrandom induction of chromosome aberrations by chemical exposures or irradiation; (4) stable positioning of individual chromosomes or chromosome regions in the interphase nucleus.

The study of the somatic association of homologous chromosomes may offer important cues to the ordered arrangement of chromatin threads in interphase nuclei: the association may be accompanied by crossing over and its genetic sequela in somatic cells. Induction of the somatic crossing over by somatic pairing of homologous chromosomes was reported from *Drosophila*, mouse and somatic human cell studies (Stern, 1936; Gruneberg, 1966; German, 1964). Homologous pairs are genetically controlled (Fedak and Helgason, 1970). In addition, evidence for the spatial relationship of chromosomes in human somatic cells, peripheral location of genetically inactive X and Y chromosomes in the interphase nucleus, and peripheral location of late-replicating chromosomes in somatic metaphases suggests that homologous chromosomes have counterpart positions in interphase nuclei (Miles, 1961; Barton et al., 1965; Prokofyeva-Belgovskaya, 1986; Ockey, 1969; Hoehn et al., 1973; Werry et al., 1977; Comings, 1980; Nagele et al., 1995; Croft et al., 1999). The chromosomal and nuclear position of a gene can influence its activity (Andrulis et al., 1998) and the position of a gene within the nucleus can be dictated by the sequences it is joined to on the chromosome (Dernburg et al., 1996).

Our results indicate that mutual positions of homologous chromosomes in human somatic cell metaphases change with age. The chromosomes tended to pair in control and to separate in senile subjects.

Table 6.1.1-2. Distances between four homologous chromosome pairs

Homologous Chromosome pairs	l^i distance	l^0		l^1		l^2 Age		l^3		l^4	
	Sex	**20-48**	**80-110**	**20-48**	**80-110**	**20-48**	**80-110**	**20-48**	**80-110**	**20-48**	**80-110**
	F	36(27)	27(27)	22(27)	35(27)	25(27)	22(27)	13 (14,5)	12 (14,5)	4,2	3,7
1-2	M	39(27)	30(27)	27(27)	29(27)	20(27)	28(27)	10 (14,5)	9(14,5)	8,6	2,6
	F,M	75(54)	57(54)	49(54)	64(54)	45(54)	50(54)	23(29)	21(29)	11,4	4,5
	F	30(27)	31(27)	27(27)	18(27)	29(27)	25(27)	10 (14,5)	22 (14,5)	1,9	7,6
3-4	M	30(27)	22(27)	30(27)	23(27)	23(27)	35(27)	13 (14,5)	16 (14,5)	1,4	4,5
	F,M	60(54)	53(54)	57(54)	41(54)	52(54)	60(54)	23(29)	38(29)	2,2	6,6
	F	33(27)	32(27)	27(27)	26(27)	23(27)	29(27)	13 (14,5)	9(14,5)	2,1	3,2
5-6	M	32(27)	27(27)	24(27)	32(27)	27(27)	29(27)	13 (14,5)	8(14,5)	1,4	4,0
	F,M	65(54)	59(54)	51(54)	58(54)	50(54)	58(54)	26(29)	17(29)	3,0	6,0
	F	40(27)	28(27)	21(27)	30(27)	25(27)	21(27)	10 (14,5)	17 (14,5)	9,1	2,1
7-8	M	38(27)	20(27)	21(27)	27(27)	20(27)	30(27)	17 (14,5)	13 (14,5)	8,1	5,0
	F,M	78(54)	48(54)	42(54)	57(54)	45(54)	51(54)	27(29)	30(29)	15,0	1,0
	F	139 (108)	118 (108)	97 (108)	109 (108)	102 (108)	97 (108)	46(58)	60(58)	12,8	2,1
Total	M	139 (108)	99(108)	102 (108)	111 (108)	90 (108)	128 (108)	53(58)	42(58)	12,7	7,0
	F,M	278 (216)	217 (216)	199 (216)	220 (216)	192 (216)	225 (216)	99 (116,5)	106 (116)	24,3	1,3

Most of studies report a tendency of homologous chromosomes to associate. Schneiderman and Smith (1962) tested 30 metaphases of skin and lymphocyte cultures of normal men and women to find that the tendency was stronger in chromosomes 3 and 16 and weaker in chromosomes 1 and 2. Pairing of homologous chromosome 1 was also seen in metaphases of female peripheral lymphocytes in vitro (Barton et al., 1963; Hens et al., 1975). Our findings in persons of 20 to 48 years support this evidence. Thus, a cumulative sample from this group showed a significant deviation from the random distribution hypothesis in favor of the proximity of homologous chromosome pairs 1 and 16. In contrast, most of homolog separation was identified in pairs 1 and 3 in subjects aged 80 to 114 (in men and women), pair 16 in women and pair 2 in men.

The change in positions of homologous chromosomes was documented in mutagen exposures. Kirsch-Volders et al. (1978) reported that exposure to depomedroxyprogesterone was associated with the change of distances between homologs of pairs 16, 18, 13-15, and 21-22. Increased distances between homologous chromosomes and centromeres and metaphase centers were seen in lymphocytes from workers exposed to phenylmercuric acetate (Verschaeve et al., 1978).

Cell differentiation and gene activity may be modified by altering positions of chromosomes relative to each other and the nuclear membrane during the cell cycle in ontogenesis as well (Heslop-Harrison and Bennett, 1984).

Differentiation and aging of human cells are attended by progressive asynchronous heterochromatinization of euchromatic chromosome regions (Beerman, 1956; Lezhava, 1980, 1999, 2001a). Chromosome conjugation is blocked by this event (Prokofyeva-Belgovskaya, 1986).

Overall, heterochromatinized chromosome regions with advancing age appear to produce the asynchrony of chromosome modification during the cell cycle. Heterochromatinization either inhibits approximation of identical regions in homologous pairs or inactivates certain genes responsible for somatic association. It is feasible that increasing asynchrony is a basis of impeded homolog alignment in some subjects at 80 years and older.

Conclusion

The statistical assessment of spatial relations of homologous chromosome pairs 1, 2, 3 and 16 in peripheral lymphocytes from subjects aged 20 to 48 years demonstrated the approximation of homologous chromosomes of pairs 1 and 16, a finding at variance with the random distribution hypothesis. In subjects of 80 to 114 years, homologous chromosomes were notably separated in pairs 1 and 3 (in men and women), 16 (in women), and 2 (in men).

CHAPTER VII

DEHETEROCHROMATINIZATION OF CHROMOSOMES

7.1. PEPTIDE BIOREGULATORS INDUCED REACTIVATION OF CHROMATIN

Aging is defined as a manifestation of complex changes in genetic processes that lead to the gradual functional disorders giving rise to senile diseases resulting in inevitable death of an organism. Hence it appears necessary to invent new medical preparations destined for slowing down 'the biological clock' and preventing senile pathologies. Special interest is being paid to peptide bioregulators - preparations of a new type (Epitalon, Livagen, Vilon), which are being successfully applied in gerontological (Epitalon, Livagen, Vilon) and geriatric practice. As a result, peptide bioregulators decrease the risk of premature aging, have an antitumor activity and stimulate functioning of the immune system regulated through the genes in chromatin domains (Khavinson et al., 2002).

It is well established that chromatin is composed of distinct functional domains. Let us remind the reader that shortly after mitosis, the chromosomal domains decondense and are repositioned in the nucleus, where they are designated as either euchromatin or heterochromatin. Heterochromatin represents the chromatin that remains condensed throughout the cell cycle except during its replication, which occurs late in the S phase. Heterochromatin includes constitutive heterochromatin, that is almost entirely composed of noncoding, tandemly repeated, satellite DNA sequences (C-bands are designated on metaphase chromosomes and are mostly localized at or adjacent to centromeric regions) and facultative heterochromatin (heterochromatinizated chromosome region), that mainly consists of "closed" potentially transcribable genes. Euchromatin, decondensed chromatin regions are early replicating and actively transcribed in almost all cells (Cremer et aL2000; Carvalho et al, 2001).

That presence of "active genes" is not enough for transcription, but existence of "active chromatin" is necessary (Lundgren et al 2000; Claussen et al, 2002). It has been suggested that progressive heterochromatinization - condensation of eu- and heterochromatic regions of chromosomes accompanied by genes inactivation occur with aging (Lezhava, 2001a, b).

In this view, we considered it urgent to determine whether the system of chromatin domains in cultured lymphocytes from old persons undergoes changes when exposed to the peptide bioregulators - Epitalon, Livagen, and Vilon. In particular, our aim was to study the variability of the levels of chromatin decondensation (dehetero-chromatinization) in: total heterochromatin; nucleolus organizer regions (activating the processes of synthesis); structural heterochromatin; and facultative heterochromatin (condensed euchromalin).

We studied donor chromosomes in 190 lymphocyte cultures obtained from 95 healthy subjects of 75-88 years of age and 50 cultures from 25 young subjects of 26-35 years of age. Two cultures (intact and bioregulator-treated) from each donor were studied to compare the indices of the affected cultures to their own control values. Bioregulators were introduced into the lymphocyte cultures at concentrations of 0.005 - 0.01 μg/ml, which had no mutagenic effect.

Description of Bioregulators

Epitalon – tetrapeptide (Ala-Glu-Asp-Gly) was obtained by targeted chemical synthesis on the basis of the amino acid analysis of complex pineal peptide bioregulator Epithalamin. Epitalon revealed a pronounced antioxidant activity; it stimulated synthetic and reparative processes, enhanced the organism resistance to stress impacts and increased average life span and reproductive period (Khavinson, 2002).

Livagen - tetrapeptide (Lys-Glu-Asp-Ala) was obtained by directed chemical synthesis on the basis of amino acid analysis of the complex preparation of the liver. Livagen increases the average level of protein synthesis in aging of animals, renovates liver proteins and induces the activation of protein synthesis in hepatocytes from old rats (Brodsky et al., 2001)

Vilon - dipeptide (Lys-Glu) was obtained by directed chemical synthesis on the basis of amino acid analysis of the complex preparation of thymus - thymalin. Vilon stimulates lowering for the risk of premature aging, has an antitumor activity and stimulates functioning of the immune system and reparative processes, strengthens the resistance of organisms to stress activities, favours prolongation of the average life span (Khavinson, 2002).

The methods of DSM, SCE and activity of ribosomal genes were described before.

Polymorphism of structural C-heterochromatin. Exhibition of structural C-heterochromatin was performed using the method of Fernandez et al (2002). The system of classification proposed by Patil and Lubs (1977) was used for comparative analysis of C-stained chromosomes in intact and bioregulatortreated cells. The C-segments of chromosomes 1, 9 and 16 were compared to the short arm of chromosome 16. According to this system of classifications results were distributed to 5 variants: a, b, c, d and e. The χ^2 value was calculated by Zaks' formula (1976).

7.1.1. Denaturation of Total Heterochromatin

Heat absorption curves, corresponding to denaturation processes in PHA-unstimulated intact leukocytes cultures and peptide bioregulatortreated cultures (Epitalon, Livagen, Vilon) showed rather complex profiles. It was mentioned that chromatin inside the nuclei denatures at temperatures about 60°C, 76°C, 88°C, and 105°C (Almagor and Cole, 1989a; Cavazza et al, 1991, Cardellini et al. 2000). Therefore, we assumed that the endotherms I, II, III corresponded to the denaturation of chromatin inside lymphocytes. Adding of peptide bioregulators to the leukocyte culture caused alterations in profiles of the heat absorption curves. In particular, adding of the peptide bioregulator Vilon (Fig.7.1.1-1) (Lezhava et al., 2004) shifted endotherms II and III to lower temperatures by 2.9^0 and 1.0^0 C accordingly. Besides, the heat redistribution was observed within stages II and III; at stage II the heat increased, but at stage III it went down. Based on previous data (Cavazza et al., 1991; Cardellini et al., 2000), we suggested that stage II of the endotherm transition was related to unfolding of 10 nm and 30 nm fibers, and stage III – to unfolding of the loops consisting of 30 nm fibers attached to the nuclear matrix.

The results indicated that the treatment of cells with Vilon induced heat redistribution between stages II and III that should be attributed to the partial (deheterochromatinization) decondensation of loops up to the 30 nm fibers. A similar picture could be seen while analysing the heat absorption curves for bioregulators Epitalon, Livagen and anticancer drug effect, i.e. in this case endotherms II, III and IV were shifted to lower temperatures as well (Almagor and Cole,1989 b; Khavinson et al., 2002,2003).

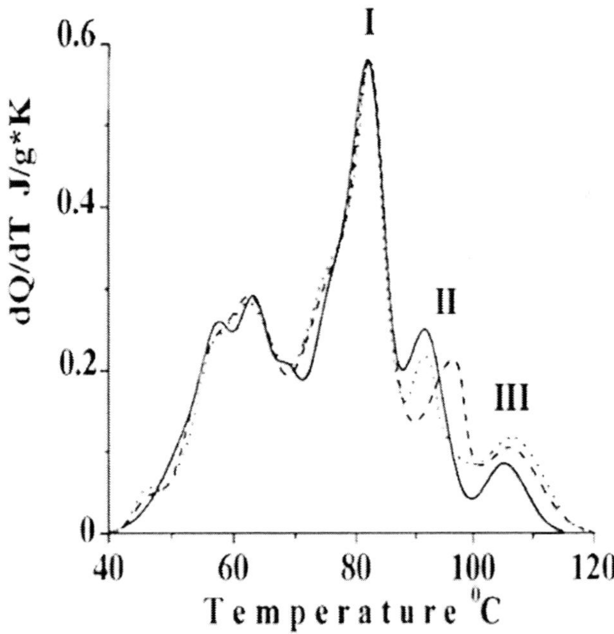

Fig. 7.1.1-1. Chromatin heat absorption capacities in lymphocyte cultures - - - from old donors ; — Vilon-treated lymphocyte cultures from old donors; .... from young donors

Thus, we can draw the conclusion that peptide bioregulators Epitalon, Livagen, and Vilon unfold the highest levels of chromatin organization, i.e. deheterochromatinize total (facultative and structural) chromatin in intact cells of old individuals (see Table 7.1-1).

7.1.2 Transcriptional Activity of Ribosomal Genes

The method of differential staining (Ag-banding) revealed that human ribosomal genes are localized in secondary constrictions (Nucleolus Organizer Regions, NORs) - in satellite stalks. The ribosomal genes take part in one of the key cellular processes - protein synthesis (Olson et. al., 2000). It was revealed that silver staining is characteristic only for the NOR intensively functioning at a previous interphase, and the staining intensity corresponds to the intensity of its functioning (Stitou et al., 2000; Lezhava, 2001a).

The ability of acrocentric chromosomes to be connected - to form associations, is determined by the presence of two chromatid satellite stalks (Lezhava et al., 1999). The associative activity of the strands positively correlates with the intensity of Ag-staining that, in turn, depends on the activity of the ribosomal genes located in NORs. The absence of satellite stalks or silver staining (caused by condensation of the stalks) also testifies to the inactivation of ribosomal genes (Hens et al., 1980; Trere, 2000).

The data obtained from the analysis of Ag-positive NORs in cultured lymphocytes derived from old donors intact and treated with bioregulators are given in Table 7.1-1. It was shown that all bioregulators strongly increased the amount of Ag-positive NORs in all acrocentric chromosomes either involved in associations or not in comparison with intact cells ($p < 0.001$). In particular, the number of Ag-positive NORs of acrocentric chromosomes involved in associations corresponded to 2.32 per Epitalon-treated cell, 2.49 per Livagen-treated cell, 2.39 per Vilon-treated cell that is significantly higher than the corresponding index for intact cultures (see Table 7.1-1). All the five bioregulators also stimulate ascending of associative activity of acrocentric chromosomes. The frequency of bioregulator-treated cells of aged individuals containing associations reliably exceeds the control value for intact cultures ($p < 0.001$). It should be noted that all the five bioregulators caused equal increases of all types of associations-DD, DG and GG. (Fig. 7.1.2-1).

Our results are in accordance with the previous data. In particular, hormones, various growth factors and chemicals induce chromosome decondensation (in old age as well) resulting in increased transcriptional activity of nucleolar organizers (Mamaev and Mamaeva, 1992; Bablishvili, 2002; Khavinson et al.2002, 2003; Lezhava et al., 2004). It was established that the frequencies of Ag-positive NORs and associations depended on the degree of condensation (heterochromatinization) of satellite strands. The chromosomes of D and G groups with well-defined decondensed satellite stalks show a strong tendency to form associations (Lezhava 2001a, b; Stitou et al., 2002). Our data support this statement.

An increase in amount and size of Ag-positive NORs, as well as in the number of acrocentric chromosomes involved in associations, in the cultures obtained from old individuals treated with bioregulators indicated deheterochromatinization of satellite stalks, as compared to control values. This could lead to the intensification of protein synthesis due to the activation of ribosomal genes in aged individuals.

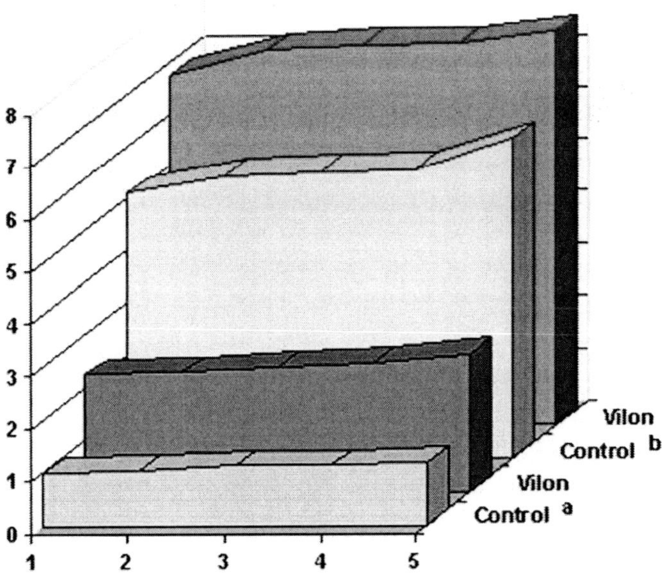

Fig. 7.1.2-1. Frequency of acrocentric chromosome associations and Ag-positive NORs in Vilon-treated lymphocyte cultures of aged individuals. a - frequency of acrocentric chromosome associations in Vilon-treated lymphocytes and control cultures; b -frequency of Ag-positive NORs in Vilon-treated lymphocytes and control cultures. On X axes – studied individuals. On Y axes – %

Table 7.1-1. Influence of peptide bioregulators (Epitalon, Livagen, Vilon) on reactivation of chromatin from old people.

Experimental conditions	Association of acrocentric chromosomes per cell	Facultative heterochromatin (SCE per cell)	Total heterochromatin	Structural heterochromatin (C-bands) Chromosomes		
				1	9	16
Control (20-40 yr.)	1.33 ± 0.06	7.7 ± 0.4	Stable conditions	Stable condition	Stable condition	Stable condition
Control (75-88 yr.)	1.17 ± 0.05	5.9 ± 0.2	Heterochromatinized	Hetero chromatinized	Stable condition	Stable condition
Epitalon	2.32 ± 0.12	8.4 ± 0.5	Deheterochromatinized	Dehetero chromatinized	Dehetero chromatinized	Stable condition
Livagen	2.49 ± 0.14	9.2 ± 0.4	Deheterochromatinized	Dehetero chromatinized	Dehetero chromatinized	Stable condition
Vilon	2.39 ± 0.11	9.9 ± 0.6	Deheterochromatinized	Hetero chromatinized	Stable condition	Stable condition

7.1.3. Heteromorphism of Structural C-Heterochromatin

The data on heteromorphism of structural heterochromatin (C- segments) in intact and bioregulator-treated (Epitalon, Livagen, Vilon) lymphocytes of old individuals for chromosomes 1, 9 and 16 are presented separately in Table 7.1-1.

The data reflecting variability of large (d and e) and small (a and b) C-segment variant frequencies in separate chromosomes appeared to be unequal for each of the 3 tested bioregulators (Epitalon, Livagen, Vilon). In particular, chromosomes 1 and 9 appeared to be heteromorphic in Epitalon and Livagen treated cells. The rate of heteromorphism (the decrease of large bands in size) for appointed chromosomes was significant ($p<0.01$). The difference from the contol indices was not noticed for chromosome 16 ($p>0.05$).

As for Vilon-treated cells (see Table 7.1-1.), large and small C-segment variants for chromosomes 1, 9 and 16 were registered with approximately the same frequency in intact cells – differences between the indices compared were not significant ($p<0.05$). It should be noted that the distribution of C-segment variants for chromosome 1, 9 and 16 remained stable and did not differ from the corresponding intact cells ($\chi^2_4=4.99$, $p>0.05$, $\chi^2_3 = 3.03$, $p> 0.05$, $\chi^2_2 = 1.07$, $p<0.05$ - for chromosome 1, 9 and 16 respectively).

The previous data showed that chromosomes 1 and 9 were characterised by the expressed variability of absolute and relative size of C-heterochromatin in some pathologies and when treated with chemicals, whereas the distribution of heterochromatin C-segment variants for chromosome 16 remained stable ($p<0.001$)(Bablishvili, 2002 ; Lezhava and Bablishvili, 2003).

According to the data obtained, bioregulator-treated lymphocytes promote a decrease of C-band size in chromosomes 1 and 9 (Epitalon, Livagen), that indicates decondensation (deheterochromatinization) of structural heterochromatin.

7.1.4. Variability of Facultative Heterochromatin (Condensed Euchromatin Regions) Based on the SCE Test

The effect of bioregulators (Epitalon, Livagen and Vilon) on SCE frequency in lymphocyte cultures of aged individuals is presented in Table 7.1-1. All the bioregulators studied caused an increase in SCE frequency in old people cells (see Table 7.1-1). The difference between two age groups: old individuals and control one was significant for all variants ($p< 0.001$).

To reveal the chromosomes responsible for the increase of average SCEs, an evaluation of SCE distribution over chromosome groups was performed. The analysis showed that Livagen and Vilon induced equal increases of SCE in A, C, D, E and G groups having a statistically reliable character. No difference between the SCE indices was observed for chromosomes of the B and F groups - the effect of Vilon was not significant ($p > 0.05$). Epitalon induced a significant increase in SCE counts in A, C, D and G chromosome groups with statistic relevance. No difference between SCE indices was observed for chromosomes of the B, E and F groups, i.e. the effect of Epitalon was not significant. These data show the selective character of each bioregulator tested on certain chromosomes.

Thus, we can conclude that peptide bioregulators, due to their ability to decondensate chromatin, promoted the release of genes repressed due to age-related heterochromatinization of euchromatin regions.

Discussion

The results obtained show that peptid bioregulators Epitalon, Livagen and Vilon significantly induce reactivation, deheterochromatinization of chromatin in lymphocyte cultures of aged individuals in comparison with control.

It was revealed that the number of chromosomes with silver-stained NORs and the frequency of associations formed by Ag-positive satellite stalks of chromatids increased in old age. These parameters are known to be positively correlated (Lernid et al., 1980); they are genetically determined (Kirsch – Volders, 1980) and reflect nucleolus organizing activity (Hens et al., 1980). Some results about the age-dependent increase of age-positive NORs, frequency of acrocentric chromosome associations and activity of ribosomal cistrons have been reported (Bablishvili et al., 2002; Lezhava and Bablishvili, 2003). Our data are compatible with these findings.

The treatment of cells with peptide bioregulators Epitalon, Livagen and Vilon induced heat redistribution between stages II and III, which was connected with partial decondensation - deheterochromatinization of a loop up to 30 nm fibers. A similar conclusion was made while studying the influence of an inorganic salt of sodium (Na_2 HPO_4) on the level of heterochromatin by DSM methods (Bablishvili, 2002).

However, insignificant dissimilarities in the C-bands of chromosomes 1, 9 and 16 were observed while studying the preparations. While Epitalon and Livagen stimulated deheterochromatinization of structural heterochromatin in chromosomes 1 and 9 (variability of chromosomes 1 and 9 in pathologies corresponds to previous data – Podugolnikova and Solonichenko, 1994), Vilon could not induce changes in the size of C-heterochromatic regions of chrorosomes 1 and 9, indicating stability of structural C-heterochromatin.

The influence of the tested bioregulators on SCE frequency in certain groups of chromosomes was of a different character. In particular, an increase in the exchange number for A, B, C, D, E and G chromosome groups was induced by Livagen and Vilon. Epitalon stimulated an increase in the exchange number for the A, B, D and G chromosome groups. These data indicate different effects of each of the tested bioregulators on specific chromosomes.

According to data published elsewhere, the processes of exchange did not occur in heterochromatin or heterochromatinized chromosome regions (Hawley and Arbel, 1993; Lobov and Podgornaya, 1999). In this view, peptide bioregulators Epitalon, Livagen and Vilon promoted an increase in the incidence of SCEs indicating the deheterochromatinization (decondensation) of euchromatic regions heterochromatinized with age.

In response to a steroid hormone, transcription in this system exhibits up- and later down regulation, analogous to puff and comp brush chromosomes (Archer et al., 1994). The complete continuum, from decondensed to fully compact, “silenced” chromatin, is compatible with the conserved topological framework of the 3D zigzag of nucleosome and linker DNA (Bednar et al., 1998).

Consequently, the obtained results demonstrated that peptide bioregulators Epitalon, Livagen and Vilon induced the activation of chromatin due to their ability to modify the heterochromatin and heterochromatinized (facultative heterochromatin) chromosome regions in the cells of old individuals.

CONCLUSION

The effect of the synthetic peptide bioregulators (Epitalon, Livagen and Vilon) have been studied on structural and facultative heterochromatin of cultivated lymphocytes from old people.

The data obtained indicate that Epitalon, Livagen and Vilon: 1) activate synthetic processes caused by the reactivation of ribosomal genes as a result of deheterochromatinization (decondensation) of nucleolus organizer regions; 2) induce unrolling (deheterochromatinization) of total heterochromatin; 3) release the genes repressed due to the heterochromatinization (condensation) of euchromatic regions forming facultative heterochromatin; 4) Epitalon and Livagen induce deheterochromatinization (decondensation) of pericentromeric structural heterochromatin of 1 and 9 chromosomes. However, Vilon does not induce deheterochromatinization of pericentromeric structural heterochromatin.

These results indicate that peptide bioregulators Epitalon, Livagen and Vilon cause activation (deheterochromatinization) of lymphocyte chromatin in the cells of old individuals.

Chapter VIII

General Conclusions

8.1. Heterochromatinization of Chromosomes

Aging of human cellular systems is consistently associated with heterochromatinization—condensation of selected transcribable chromatin (chromosomes) regions and repression of genes (Prokofyeva-Belgovskaya, 1986; Lezhava, 1980,1984b, 2001a,b).

Biochemical and genetic findings accumulated over the past decade have established that the condensation of eukaryotic DNA in chromatin functions not only to constrain the genome within the boundaries of the cell nucleus but also to suppress gene activity. This genetic repression extends from the level of the nucleosome, the primary unit of cromatin organization, where coiling of DNA on the surface of the nucleosome core particle impedes access to the transcriptional apparatus, to the higher order folding of nucleosome arrays and the organization of the silent regions of chromatin. Chromatin structure is inextricably linked to transcriptional regulation, and show how chromatin is perturbed so as to facilitate transcription (Wu, 1997). With histone acetylation, this post-translation modification has been correlated with the processes of transcription and chromatin assembly. Acetylation occurs at specific lysines in the flexible N-terminal histone tails that protrude from the nucleosome surface. Hyperacetilation of histones is associated with transcriptional activity or the potential for activity, whereas histone hypoacetylation is correlated with transcriptionally silent chromatin and heterochromatin. Histone acetylation is also associated with the active deposition and maturation of newly assembled nucleosomes during DNA replication. Acetylation reduces the net positive charge of the histones and weakens interaction with the DNA, inhibits the higher order folding of nucleosome arrays and disrupts specific interactions with nonhistone regulators (Garcia-Ramirez et al., 1995; Wu, 1997). It has been suggested that acetylation of H4 during nucleosome assembly regulates the binding of H1 and the ability of chromatin to condense. While in some cases active genes are hyperacetylated and contain H1, it has also been reported that while H1 binds to acetylated oligonucleosomes, this binding inhibits transcription. In addition, studies have demonstrated that histone acetylation alters the capacity of histone H1 to condense chromatin and that the presence of H1 affects the ability of transcription factors to interact with the DNA. Recent studies have also shown that the the retinoid receptor, a receptor known to function in part by recruitment of HATs, must also recruit an activity for displacement or remodeling of the linker histone H1. These results argue that displacement of H1 is required prior to acetylation of the target gene and activation of transcription. In addition, studies involving steroid hormone receptors, also known to interact with HATs, have shown that activation involves phosphorylation of H1 that results in a reduced affinity of H1 for chromatin (Herrera et al., 2000).

A survey of experimental evidence on age-dependent alteration of gene expression in laboratory animals (Kanungo, 1984) has shown that the activity of several enzymes decreases with advancing age and may be improved with steroid hormones. Populations of receptors and their affinity for hormones are reduced by aging, with a decline in gene expression. The alteration in gene expression is indicated by a change in primary isoversions of many complex enzymes. Treatment of postmitotic mammalian cell chromatin with DNAase and micrococcal nuclease demonstrated that heterochromatinization of chromatin is increasingly greater with age, causing the age-dependent depression of gene expression. Analysis of products of micrococcal nuclease-induced DNA hydrolysis with circular dichronism and electrophoretic techniques has shown that chromatin is increasingly condensed with age both in an unfolded nucleosomal chain and in supranucleosomal structures (Konoplya et al., 1988). Furthermore, senility is associated with a tighter chromatin packing by induced Ca, Mg-dependent endonuclease at the nucleosome level and a lower availability of linker DNA for nuclease (Dubrovsky and Berdyshev, 1986). Chromatin-melting temperature is higher in older age, signifying higher rates of heterochromatinization (Lezhava et al., 1993).

The mechanisms of chromosome heterochromatinization are unknown. It was inferred that heterochromatin has genes for heterochromatinization of chromosomal euchromatin regions in *Drosophila* and other organisms. In humans and mice, heterochromatinization of the X chromosome is determined by loci of its centromeric heterochromatin (Birstein, 1976). DNA repetitions may provoke heterochromatinization of chromosomes (Foural et al., 2002).

Chromosome condensation in the G2 and mitotic periods was assumed to depend on chromomeres that mediate the formation of disulfide bonds between molecules of certain chromosomal proteins (Sumner, 1976; Ronne, 1980). The chromomeres may control the gradual heterochromatinization of chromosomes in lymphocytes cultured from senescent humans (Ronne, 1980).

A molecular basis of heterochromatinization and sex differentiation was examined in mealy bug using a polysterine sulfonate histone binding test and autoradiography (Berlowitz, 1974). It was found out that arginine-rich histones have a major functional involvement in the transcription, while lysine-rich histones provide for a strong condensation of heterochromatin. It is believed that chromosome condensation and decondensation are basically dependent on two factors: the number of histone globules per unit of DNA length and the condition of DNA at the time of association or dissociation of the H1 histone, an inducer of chromosome condensation; it implies that the chromatin structure is directly related to the histones, although the mode of their relationship with DNA is determined by nonhistone proteins (during chromosome decondensation genes are activated by nonhistone regulator proteins) (Khesin and Leibovich, 1976). It should be stated, however, that senescence alters histone fractions: the quantity of the H1 histone is substantially increased and that of lysine-rich H2A and H2B histones is moderately increased, quantities of the arginine-rich H3 and H4 histones probably decrease. Being natural polycations, the histones are capable of both specific and nonspecific interaction with chromatin anions and polyanions: derepression of transcriptionally active DNA, denuded DNA sites, RNA polymerase at nucleosome surfaces, regulator and other nonhistone proteins. The overall effect is inhibition of the transcription (Kucherenko et al., 1983). It is known that in old age DNA preferentially binds to the arginine-rich histones H3 and H4, making an especially strong bond with H3 (Hahn et al., 1969). It may explain the fact that these histones inhibit RNA and protein synthesis stronger than lysine-rich histones. It is H3 and H4 that may induce the inactive, condensed status of chromatin (Das et al, 1964; Mirsky et al., 1972). These properties of the arginine-rich histones may be related to their greater ability of covalent modifications like acetylation and methylation, as compared to other histones.

Acetylation mediates the transcription, thereby decreasing histone charge densities. This impairs the DNA-histone relationship and enables the RNA polymerase to attach to DNA. It is confirmed by evidence that histone acetylation rates during the mitotic cycle are highest at the interphase and lowest at the pro- and metaphase, when chromatin is most condensed (D'Anna et al., 1977; Kucherenko et al., 1983). In addition, it is feasible that old age reduces H2A and H2B methylation rates (promoting chromatin decondensation and transcription) and increases H3 and H4 methylation (with a higher chromatin condensation). The increase in histone methylation with age may occur because of altered rates of methylase or dimethylase phosphorylation since in vivo and in vitro experiments have identified the ability of methyltransferase III (which methylates primarily arginine-rich histones) to phosphorylate (Kucherenko, 1983; Catania and Fairweather, 1991).

Experimental evidence indicates that histone H1 phosphorylation rates significantly decline with age (Kanungo, 1982). The loss of stretching ability by contractile nonhistone proteins is thought to have a major role in the age-related irreversible chromosome condensation (Prokofyeva-Belgovskaya, 1986). Importantly, the tubulin-like protein is phosphorylated by AMP-dependent protein kinase; moreover, it has a protein kinase activity toward the histones and its own molecules (this activity regulates the protein's contractility). This enzymic activity decreases sevenfold with aging (Schroder et al., 1983; Howard, 1996).

The rates of nonhistone protein phosphorylation are highest at the mid-G1-phase and early S-phase and lowest at the G2-phase and mitosis. Therefore, the reported aging-associated decrease in in vivo and in vitro rates of phosphorylation is an expected finding (Karn et al., 1974; Kanungo, 1982).

Another important determinant of chromatin function in old age is cations. Aluminum and copper accumulates in tissues of old humans and animals, and presumably acts as a binding agent which mediates euchromatin condensation (Jokhadze and Lezhava, 1994; Kawanishi et al., 2002).

Extensive earlier investigations have demonstrated that histone H1 facilitates higher order chromatin packing and confers insolubility even to mononucleosomes at physiological ionic strengths (Hansen, 2001). The globular domain of histone H1 binds to the major groove of DNA, whereas core histones bind to the minor groove. Native nucleosomal DNA wrapping through core histone interactions therefore seems to be required for nucleosomes to assemble into higher order structures that lead to chromatin condensation. This wrapping is known to neutralize positively charged amino acids by a close association with negatively charged DNA phosphate groups (Wildak et al., 2002)

In interphase nuclei, heterochromatic regions are often associated with specific chromosomal proteins, post-translational modifications, and methylated DNA (Khochbin , 2001). In mitotic cells, chromosome condensation requires post-translational modifications and the action of an ATP-dependent complex called "condensin" to introduce positive DNA supercoils into DNA substrates in the presence of topoisomerases (Uhlmann, 2001). ATP-dependent chromatin remodeling machines of the SWI/SWF family are involved in many cellular processes in eukaryotic nuclei, such as transcription, replication, repair and recombination (Langst and Becker, 2001).

The heterochromatin protein 1 (HP1) immunostaining on polytene chromosomes from Drosophila larval salivary glands was used to show enrichment of the protein in pericentric heterochromatin. The HP1 homologues have been found in species ranging from fission yeast to humans where it is associated with gene silencing. It has been generally assumed that these sites might constitute euchromatic sites of transcription repression by HP1. Indeed, several genes located at one of these sites have increased transcription levels in mutants for HP1 (Kellum, 2003).

Sir2p is an NAD (nicotinamide adenine dinucleotide)–dependent histon deacetylases that participates in the formation of compacted chromatin in budding yeast. In Saccharomyces cerevisiae, there are four proteins homologous to Sir2p (1-4). Although some of these proteins can affect silencing to different extents, their function is in general not clear. In humans there are at least seven Sir2-like proteins (SirT 1-7). SirT1 is a human histone deacetylase with preference for H4-K16 (Gartenberg, 2000).

Vaquero et al (2004) demonstrated that SirT1–mediated formation of repressive chromatin is mediated formation through at least four SirT1–coordinated events. First, SirT1 preferentially catalyzes the deacetylation of H4-K16. Previous reports have shown that acetylated H4-K16 has a "chromatin mark" association with euchromatin and that heterochromatin is hypoacetylated in K16. Secondly, SirT1 interacts with histone H1, and this interaction recruits histone H1 to establish repressive chromatin. Moreover, H1 has recently been found to be recruited to the MyoD gene by the transcription repressor Msx-1 involved in muscle differentiation (Lee et al., 2004). Thirdly, SirT1 catalyzes deacetylation of H1. It is possible that since H1 is involved in formation of compacted chromatin structures in part by its interaction with linker DNA, its deacetylation may produce a similar effect as that shown for core histones. Thus, deacetylation of both core histones and H1 might produce a stronger effect on chromatin compaction that either separately. Fourthly, targeted to a reporter integrated into euchromatic regions of the mammalian genome resulted in the reduction of methylated histone H3-K79 at the promoter and within the coding region. SirT1 interacts with some transcription factors.

These factors likely recruit SirT1 to specific genes to repress transcription. Finally, aging produces a general heterochromatinization, a decrease in repair of chromatin aberration (Lezhava, 2001, Lezhava et al., 2004), an increase in H4-K20 trimethylation (Schotta et al., 2004), and telomere shortening. In other words, aging is associated, in part, from failure of the proper control of chromatin structure (Vaquero et al., 2003, 2004).

Evidence on aging-related chromatin (chromosome) alteration suggests that progressive heterochromatinization (euchromatic region condensation) is controlled by the genotype and environment; it is a structural consequence of a lower gene expression. The alteration of gene expression results from modifications of histone and nonhistone proteins of chromatin (phosphorylation, acetylation, methylation), with a change in numbers of covalent bonds and charges. Heterochromatinization is made persistent by chromatin intermolecular linkages that are produced by physicochemical agents and are inaccessible to repair systems because of progressive heterochromatinization.

8.2. Heterochromatinization as a Key Factor in Aging

The analysis of personal and published evidence suggests that a leading factor in aging is impairment of chromosomes' cyclical properties (progressive heterochromatinization) resulting in a less intensive repair of damaged chromosomal DNA and increased numbers of cells with chromosome aberrations.

The chromosome is a heterogeneous structure which is linearly differentiated in terms of the genetic activity, reproductive time and condensation rates. During the mitotic and meiotic cycles, it undergoes cyclical modification and controls its genetic function The structural condensation of chromosomes closely correlates with functional heterogeneity. Tightly condensed (heterochromatin, heterochromatinized) chromosome regions are genetically inert (they replicate late and are not transcribed). Decondensed (euchromatin) regions are functional; contain RNA and DNA that rapidly display labels (and have a high sensitivity to DNAase I and nuclease I. The gene activity begins with chromosome decondensation at a hypersensitive site of DNAase I (Nagl, 1985; Prokofyeva-Belgovskaya, 1986; Cremer et al., 2000; Carvalho et al., 2001).

The euchromatin and heterochromatin regions make up a heterocyclical gene system with high plasticity which is provided by labile cycles of numerous identical genes of inert regions. The inert-region genes are a powerful intracellular strategy which was evolutionarily inbuilt into the nucleus to control the cells' metabolic potential by readjusting nuclear heterocyclicity. The regional chromosomal cycles are genetically determined (Prokofyeva-Belgovskaya, 1986).

There is ample experimental evidence indicating that chromosome lesions produced by physicochemical exposures and viruses are more common in heterochromatic than euchromatic regions (Prokofyeva-Belgovskaya, 1986; Forni, 1996; Lezhava, 2001a).

Thus, it was demonstrated by induction of premature chromosome condensation that heterochromatinization in the G2 period was greater than in the G_0 and G_1 periods (Roa and Johnson, 1974; Marcus and Sperling, 1979). Chromosomes are more sensitive in the double stalk G_2 period, as compared to the one stalk G_0 and G_1 periods (Machavariani, 1976; Pincheira et al., 1993). In addition, the radiosensitivity is substantially higher during mitosis (the time of the highest heterochromatinization) than the G_2 period (Belloni et al., 1977). Mitotic cells are more sensitive to X rays than interphase cells in the S phase (Dewey et al., 1972).

The higher incidence of major damages in heterochromatin or heterochromatinized DNA domains is explained by the fact that the mutation-inducing damages can be repaired only at euchromatic DNA regions that are involved in active transcription and physically accessible to repair enzymes (Yielding, 1974; Vilenchik, 1981). This explanation is consistent with findings of an autoradiographic study of UV-induced unscheduled DNA synthesis: the amount of radioactivity recovered in euchromatin nuclear regions was two times as high as in heterochromatin ones (Harris et al., 1974). Furthermore, significantly lower DNA repair synthesis rates were observed during mitosis than the G1 phase (Giulotto et al., 1978). The deterioration of DNA repair has been reported for aging mammalian cells (including human diploid strains and lymphocytes) (Lambert et al., 1977; Hart et al., 1979; Small and Aronson, 1988; Evans and Bohr, 1994; Saadat et al., 1998; de Boeret et al., 2002).

Our studies have demonstrated a dramatic decline in unscheduled DNA synthesis rates at 80-90 years with UV-irradiation by 10-15 J/mm^2 ($P < 0.03$; $P < 0.01$) , as compared to middle adulthood.

The incidence of SCE (a mode of repair) has been investigated in vivo and in vitro in fibroblasts and lymphocytes from humans aged 60 or older and from animals of different ages. In certain spontaneous culture systems, the SCE incidence decreased with increasing donors' ages and numbers of passages (Schneider et al., 1979; Arce, 1981), while other studies reported no variability of lymphocyte chromosome SCE over a wide range of ages (Galloway and Evans, 1975; Morgan and Crossen, 1977; Hardner et al., 1982; Vormittag, 1983) or its higher incidence with age (Goh-Kongo-oo, 1981; Melaragno and Smith, 1990). According to the published data, the exchange processes do not occur in heterochromatin or heterochromatinized chromosome regions (Prokofyeva-Belgovskaya, 1986; Hawley and Arbel, 1993; Lobov and Podgornaya, 1999).

In our studies, peripheral lymphocytes from very old individuals tended to have lower SCE counts (especially in A1 and C chromosomes). The discrepancy of these findings with reported evidence may be related to the age of our subjects (80 years and older); their cells exhibited progressive heterochromatinization of selected chromosomal regions (Lezhava, 1979, 1980, 1984b), presumably reducing sister chromatid exchange rates (Hoo and Parslow, 1979; Tada-Aki and Hori, 1983; Lezhava, 2001a, b).

These facts suggest that physically blocked (heterochromatinized) chromosomal regions are more prone to damage, and DNA repair is severely impaired in old age.

The development and aging of cellular systems are associated with massive changes in the cyclical behavior of chromosomes and their ability to condensate and decondensate. Heterochromatinization unfolds in chromosomes of aging differentiated plant and animal cells (Prokofyeva-Belgovskaya, 1986).

Support for this was provided by our observations (Lezhava and Chitashvili, 1979; Lezhava, 1991). Analysis of the distribution of A group homologous chromosomes 1, 2 and 3 and group E chromosome 16 in lymphocyte metaphases showed homologous association in middle age, but not in subjects aged 80 to 110 years, suggesting a low conjugation of selected chromosomes in somatic cells of the old organism. It is likely that the onset of heterochromatinization in the chromosome regions disrupts synchrony of their cell-cycle transformations, thereby inactivating certain homologous association genes.

Since chromatin has fundamental biological roles, its alterations might explain complexities of the aging process. According to Cutler (1978), the physicochemical and functional evolution of chromatin with age occurs as increasing terminal stability, lower salt extraction of chromatin proteins, and alteration of transcription (usually repressed). The postembryogenetic program of protein and nucleic acid synthesis harbors a lot of implications for the time of aging. Repression and derepression of structurally similar loci on the DNA template dictate that the synthesis of some similar proteins be replaced by that of others. This is a basis of the complex biochemical differentiation which underlies cell and tissue aging in higher animals (Howard, 1996).

If the cellular genomic activity is differentially altered by aging, transcriptions in "juvenile" and "old cell" populations should be differential (Hahn, 1970). A twofold decrease in chromatin function has been identified in hepatic, cerebral and sarcoma cells from old C57B8/6 mice (Cutler, 1978). To elucidate in DNA methylation occurring by inappropriate epigenetic control during aging, Kang et al. (2001) compared fetal bovine fibroblasts and their aged noemycin – resistant versions using bisulfite – PCR technology. Reduction in DNA methylation was observed in euchromatin repeats. Contrastingly, a stable maintenance of DNA methylation was revealed in various heterochromatic sequences (satellite I/II/alphoid and Bov-B). These global, multi-locus analyses provide evidence on the tendency of differential epigenetic modification between genomic DNA regions during aging.

The aging of mammals, such as rats, mice, cattle, is associated with a prominent tissue-specific reduction in DNA 5-methylcytosine levels (Vanyushin and Romanenko, 1980; Mazin, 1994). The age-related DNA undermethylation was not random: it involved repetitive sequences in genome, but not unique DNA nucleotide sequences. Presumed causes of the undermethylation were lower cellular contents of CH3-groups, altered activities and specificities of DNA methylases (and probably dimethylases), and structural chromatin changes (condensation, linkages) interfering with DNA access to methylated DNA sites (Catania and Fairweather, 1991).

Our old age subjects showed increased percentages of condensed chromatin, the peak of condensed chromatin heat absorption and masses of centromeric heterochromatin qh1 (without preparation pretreatment and staining with unbuffered Unna's blue); these findings indicated progressive heterochromatinization in aging differentiated cells (Lezhava, 1984b; Lezhava and Dvalishvili, 1992; Lezhava et al., 1993). An increase of heterochromatinization with age has been documented by experimental studies in mice liver cells, in humpbacked salmons at puberty (Mozzhukhina, 1984). An experimental study on the age-related change in gene expression (Kanungo, 1984) has shown that enzyme activities decreasing with age are recoverable with steroid enzymes. The pretreatment of postmitotic cells with DNAase I and micrococcal nuclease revealed greater chromatin packing (heterochromatinization) in older animals; according to Kanungo, the alteration of gene expression associated with it is a primary cause of aging.

It should be stated that similar changes are displayed by lymphocyte genomes of very senile individuals and long-term lymphocyte cultures (120 to 144 hours) during which progressive chromosome heterochromatinization is also found (Ronne, 1980, Dvalishvili, 1983; Panno and Naiz, 1984). The association incidence decreases (satellite stalks are condensed) with time (Sigmund et al., 1979; Ardito et al., 1983; Frolov, 1986; Lezhava and Dvalishvili, 1992; Lezhava, 2001a, b).

The ribosomal gene activity is massively reduced late in a lifetime (Lezhava, 1984a), indicating repression of the total genomic transcription and specific rRNA gene repression (Cutler, 1978). The synthesis of ribosomal RNAs is depressed early in the aging process (Van-Ganzen, 1979). Biochemical and autoradiographic studies demonstrated a 50 percent decrease in the RNA synthesis and fibroblast nucleolar labeling in aging humans (Ryan and Cristofalo, 1975; Bownian et al., 1976). Ribosomal cistrones are known to be inactivated with aging in several mammalian species, including humans. The ribosomal gene change may influence the rates of aging (Gaubatz, 1976). In vivo and in vitro rates of ribosomal RNA transcription in lymphocytes from senescent individuals are significantly lower than those from middle-aged individuals and in short-term cultures (Lezhava, 1984a). This repression is thought to be induced by heterochromatinization of chromosome NOR (Van-Ganzen, 1979; Lezhava and Khmaladze, 1988b; Howard, 1996; Lezhava, 1996).

The counts of Ag-positive acrocentrics (active ribosomal genes) and the incidence of acrocentric association have been reported to decrease with age (Buys et al., 1979; Denton et al., 1981). Available evidence suggests functional similarity of the nucleolar organizer in aging humans and animals. The counts neither of Ag-positive NOR per cell (with or without chromosome association) in 80-year or older individuals were significantly below middle-age values. The quantity of Ag-positive NOR and the association incidence decrease in senile age because of heterochromatinization of satellite stalks in acrocentric chromosomes, reflecting altered synthetic activities of ribosomal cistrones (Orey, 1974; Lernid et al., 1980; Taylor and Martin-De Leon, 1981; Lezhava, 1984a; Lezhava and Khmaladze, 1989; Lezhava and Dvalishvili, 1992). Therefore, the relationships between homologs, heat absorption and mass of condensed chromatin, RNA transcriptional activity, numbers of Ag-positive acrocentric NOR and association incidence, as seen at 80 years and older, indicate that age-related changes in human somatic cells are determined by heterochromatinization (condensation) of selected euchromatin regions in chromatin threads. Moreover, the recent research has suggested that the incidence of spontaneous and induced (by chemicals or gamma-irradiation) aneuploidy and chromosome aberrations substantially rise with advancing age (Deknudt and Lionard, 1977; Lezhava and Khmaladze, 1978; 1988a; Kuznetsov et al., 1980; Esposito et al., 1989; Smith et al., 1990; Ghosh et al, 1991, 1992; Takeshita et al., 1992; Ohtaki et al., 1994; Catalan et al., 1995; Ramsey, 1995; Thierens et al., 1996; Fenech et al., 1997; Xiao et al.,1998; Bucvic et al., 2001; Lezhava, 2001a; Maeng et al., 2004; Vorobtsova et al., 2004; Warburton, 2005).

The above discussed phenomena of aging suggest a relationship between progressive heterochromatinization of chromosome regions, declining intensity of repair, and higher incidence of chromosome aberrations. Evidence that excessive heterochromatinization with aging inhibits repair enzymes, with a secondary increase in numbers of cells with chromosome aberrations, indicates a key role for heterochromatinization in the genetic mechanisms underlying the aging process. Researchers working in the field life sciences should foresee these date while seeking the ways to elongate lifespan of men and eukaryotes in general.

8.3. The Possible Mechanism of Aging Pathologies

It should be noted that progressive heterochromatinization controlled by genotype and environment is the structural result of gene expression decrease, consequently causing the lesions in DNA. In support of this suggestion the established data adduce, the presence of "active genes" is not enough for transcription, but the existence of "active chromatin" is necessary as well (Lundgren et al., 2000; Claussen et al., 2002).

Changes in gene expression are the results of modification of histone and non-histone proteins (phosphorylation, acetylation, and methylation) caused by the quantitative changes of covalent links and changes. Chromosome heterocroma-tinization results in increasing the number of somatic mutations with aging and a decline in the level of damaged repair of DNA, that causes lesions of cells structural and functional organization leading different systems of organisms to cancerous or non-cancerous diseases (Clayson et al.,1994).

Heterochromatinization progressively increasing with aging favours inactivation of a number of once functioning "active genes". It blocks certain stages of normal metabolic processes of cell systems, causing a deficit of many specific enzymes to sooner or later result in aging pathologies.

The action of genetic systems reveals general rules of behavior of such systems as a connection between the structural and functional interrelation of "directing" and "directed" structures and reverse connections. In this respect, it should be noted those heterochromatin regions in chromosomes can be reversed. A great number of physical and chemical agents, bioregulators and hormones (Kanungo, 1984; Khavinson et al., 2002, 2003; Lezhava and Bablishvili, 2003; Lezhava et al., 2004) cause their deheterochromatinization (decondensation) releasing the now inactive (once were active) genes that seems to favour purposeful treatment of diseases.

Thus, the proposed genetic mechanism (heterochromatinization) of senile pathogenesis helps to evaluate the importance of external and internal factors in the development of diseases and enables to employment of the proper tools for therapeutic treatments.

Glossary of Terms

A

Aberration. Structural change of a chromosome produced by its break; new combinations of free ends may effect rejoining.

Acentric chromosome. A chromosome without a centromere.

Acridine. A type of mutagen that intercalates between the bases of a DNA molecule and causes single-base insertions or deletions.

Acrocentric chromosome. A chromosome the centromere of which is located in close proximity to one of its ends.

'Active' chromatin. Unwound interphase chromatin.

Adenine (A). A nitrogeneous purine base found in DNA and RNA.

Anaphase. A phase in the mitotic and meiotic nuclear division with movement of chromatids (daughter chromosomes) or chromosomes toward cell poles.

Anareduplication. Spindle abnormalities at anaphase result in equatorial clustering of chromosomes and chromosome replication within interphase.

Aneuploidy. Chromosome gain or loss making an odd number relative to a haploid set.

Association of acrocentric chromosomes. Short arm-to-short arm position or conjugation of certain satellite-bearing chromosomes in human somatic cell metaphases.

Autoradiography. A process for the production of a photographic image of the distribution of a radioactive substance in a cell or large cellular molecule; the image is produced on a photographic emulsion by decay emission from the radioactive material.

Autosomes. All the chromosomes different from the sex chromosomes.

B

Barr body. A condensed, inactivated mammalian X chromosome found in certain cells; stains dark during interphase.

Base. Single-ring (pyrimidine) or double-ring (purine) component of a nucleic acid.

Base pair. A pair of nitrogenous bases, most commonly one purine and one pyrimidine, held together by hydrogen bonds in a double-stranded region of a nucleic acid molecule, commonly abbreviated bp.; the expression is often used interchangeably with the expression of nucleotide pair.

C

Cell cycle. The growth cycle of an individual cell; in eucaryots, the cell cycle is divided into G_1, S (DNA synthesis), G_2 and M (mitosis).

Centromere. The region of the chromosome associated with spindle fibers and participating in normal chromosome movement in mitosis and meiosis.

Chicago Conference. To develop a unified standard system, the 3rd International Congress on human genetics took place in the university of Chicago in 1996, where a new nomenclature system was adopted to describe the human chromosome set and its anomalies.

Chromatids. The longitudinal subunits produced by chromosome replication and joined at the centromere.

Chromatin. The aggregate of DNA and histone proteins that make up an eucaryotic chromosome.

Chromocenter. The aggregate of centromeres and adjacent heterochromatin in the nuclei of *Drosophila* larval salivary gland cells.

Chromomeres. Microscopically visualizable chromosomal structural units of definitive size and positioning that are related by achromatin threads.

Chromosomes. Carriers of genes, the determinants of heredity. Self-reproductive nuclear structures containing DNA, DNA-bound histones and other proteins, RNA, lipids, Ca and Mg ions.

Colchicine. A chemical that prevents formation of the spindle during nuclear division.

Conjugation (synapsis). Pairing of homologous chromosomes.

Crossingover. Exchange of counterpart sites between chromatids of homologous chromosomes, usually resulting in chiasm formation and gene recombination.

Cytogenetics. The discipline concerned with the genetic implications of chromosome structure and behavior.

Cytological map. Diagrammatic representation of a chromosome.

Cytoplasmic inheritance. Transmission of hereditary trait through self-replicating factors in the cytoplasm; for example, mitochondria and chroloplasts.

Cytosine (C). A nitrogenous pyrimidine base found in DNA and RNA.

D

Dark repair. Any type of DNA lesion repair requiring no cell exposure to visible light. Two known mechanisms of dark repair are (a) excision of DNA lesion and resynthesis; (b) postreplication recovery which appears to be recombination and replication of damaged DNA after the first replication cycle.

Deletion. Elimination of chromosome regions by loss of large or small chromosome fragments carrying genetic information. Deletions may occur at various points. Deletion types vary with break numbers and locations. If a gene-bearing chromosome end is deleted by a break in a chromosome arm, the arm is shortened. Simultaneous breaks in two arms eliminate both chromosome ends; however, the open (acentric fragment-deprived) ends may rejoin to produce a ring chromosome during the mitosis. Deletions may occur because of two simultaneous breaks within a chromosome. Break sites fuse, and the chromosome is contracted, with elimination of an internal chromosomal region (interstitial deletion). Deletions affect gene arrangement and relationship.

Denver Conference, sponsored by the American Cancer Society, was held in 1960 at the University of Colorado. It first discussed standards for a human chromosome nomenclature.

Deoxyribonucleic acid (DNA). A fundamental part of the cell nucleus. Molecular weight 10^6-10^{11}. DNA is made of helical polynucleotide chains. Nucleotides contain one of heterocyclical bases (adenine, guanine, thymine, cytosine, 5-methylcytosine or, less commonly, 5-hydroxymethylcytosine, e.g. in bacteriophages), sugar, deoxyribose and a phosphoric acid residue, all bonded. The Watson and Crick definition describes the DNA structure as two 20Å-thick polynucleotide chains that spiral rightward around a common axis. The chains are counterdirected: phosphoester bonds 3'-5' have the 5'→3' path in one chain and 3'←5' in the other. Purine and pyrimidine bases are directed inward at 90° to the intrahelical axis. Bases have a complementary pattern in both chains: purines of one chain are hydrogen-bonded to pyrimidines of the other. Structurally it means that adenine is always bonded to thymine and guanine to cytosine. Base pairs have a 3.4Å spacing; one loop encompasses 10 base pairs.

Dimer. A protein formed by the joining of tow polypeptide subunits.

Diploid. Refers to an organism cell with two complete sets of homologous chromosomes.

Diplotene. A phase in meiotic cell division, when conjugating homologous chromosomes begin contraction and longitudinal splitting into two recognizable units (chromatids). Pairs of bivalent chromatids diverge, but the bivalent is preserved since exchange-produced chiasms hold chromosomes together.

Dizigotic twins. Twins that result from the fertilization of separate ova; genetically related as siblings; also called fraternal twins.

DNA methylase. An enzyme that adds CH_1 groups to certain bases, particularly cytosine.

DNA repair. Any of several different processes for restoration of the correct base sequence of a DNA molecule into which incorrect bases have been incorporated or whose bases have been modified in some way.

DNA replication. The copying of a DNA molecule.

Domain. A folded region of a polypeptide chain that is spatially somewhat isolated from other folded regions.

Down syndrome. The clinical features of the karyotyoe 47, +21 (trisome 21).

Duplication. Multiplication of identical chromosome regions resulting in phenotypical change. A chromosome region may be duplicated because of unequal chiasm formation. If genes of a normal chromosome have an ABC order, a gene duplication may induce ABBC or ABBBC state.

E

Endomitosis. An independent division increasing the number of chromosomes during nuclear growth.

Endonuclease. A division-independent increase in the number of chromosomes during nuclear growth.

Endoreduplication. Refers to the chromosomes of the cells that enter the mitosis but do not divide into two daughter cells. The chromosomes return into interphase with nonseparated sister chromatids. Diplo- and quadrichromosomes occur in prophase.

Euchromatin. A genetically active part of chromatin that differs from heterochromatin by the absence of condensation in interphase.

G

Gene. A region in a nucleic acid molecule. A gene-specific nucleotide sequence makes it a discrete unit of function. The gene may be altered by mutation.

Genes 5s, 18s and **28s** include repeats of nucleotide sequences (tens, hundreds, rarely thousands). These genes are multiple (100 to 1000; in humans 300 to 2000) and reside in centromere-adjacent heterochromatin regions of one or several chromosomes (in human D, G group and A1 chromosomes).

Genome. A complete set of genes contained by a haploid chromosome number. The genome is a genetic entity.

Group 4-5. Large chromosomes with submedian centromeres. The chromosomes show scarce difference, although chromosome 4 is somewhat longer.

H

Heterochromatin (constitutive). Intensely staining compact chromosome regions that are always present in both homologous chromosomes throughout the cell cycle. Constitutive heterochromatin is generally genetically inert. It mediates expression of Mendelian traits, replicates lately, produces chromocenters at interphase, and is more liable to breaks. Contains highly repetitive DNA nucleotide sequences and regulates chiasm formation during meiosis.

Heterochromatin (facultative). Occurs in one of homologous chromosomes due to euchromatin region spiralization. It replicates lately and is not despiralized at interphase.

Heterochromatinization. Condensation of transcribable euchromatin regions of chromosomes.

Homologous chromosomes. Structurally identical chromosomes with identical linear positioning of counterpart loci.

Hyperdiploidy. Gain of an odd number of chromosomes relative to a haploid set.

Hypodiploidy. Loss of an odd number of chromosomes relative to a haploid set.

I

Interphase. A cell stage between two divisions.

Inversion. Change of gene arrangement order in a chromosome produced by 180° rotation of an intrachromosomal region because of two breaks. If both breaks occur within one chromosome arm, inversion is defined as paracentric. If breaks occur on both sides of the centromere, inversion is pericentric.

In vitro /L/. In glass. Refers to experimental growing of cell in culture.

In vivo /L/. In living tissue.

Isochromatid break. A break involving the same site in both chromatids.

K

Karyotype. A diploid set of chromosomes classified according to size, arrangement, and number.

L

Locus. A chromosome region corresponding to gene; a homologous chromosome region occupied by identical genes (alleles).

London Conference (1963). The second conference on standardization of a normal human karyotype. It was proposed that identification and characterization of human chromosomes include important features like secondary constrictions and autoradiographic patterns of H3-labeled thymidine.

M

Meiosis. During development sex cells undergo meiotic division which reduces twofold chromosome numbers in cells. Meiosis normally includes two successive nuclear divisions: reduction (leaving half the number of chromosomes and making diploid cells haploid ones) and equation (with a haploid set retained by cells).

Mendel's laws. Patterns of hereditary factor distribution discovered by Mendel (1865).

1. Law of homogeneity. First-generation hybrids (F1) from genetically pure (homozygous) parents are homogeneous.
2. Law of fission. Second-generation individuals (F2) resulting from auto-fertilization or matings, such as brothers and sisters of F1 individuals, display exact progenitors' (P) traits. If a trait of one parent dominates over a (recessive) trait of the other, fission in F2 is 3:1, that is, 75 percent of individuals derive their phenotypes from parents with a dominant trait and 25 percent from parents with a recessive trait.

3. Law of independent assortment, or new gene combination, involves the offsprings of the hybrids whose parents were dissimilar in more than one pair of traits. In this offspring, not a combination of parental traits or genes, but each pair of traits obeys the fission law.

Metareduplication, or C mitosis. A metaphase event. Spindle abnormality at this phase is associated with chaotic dispersal of chromosomes in the cytoplasm.

Mitosis. A mode of nuclear division, also known as indirect nuclear division, or karyokinesis. Mitosis is a regular distribution between daughter nuclei of chromatids produced by chromosome reduplication and gene or genetic material transmission from one cell generation to another. The two daughter nuclei resulting from mitosis are usually genetically identical.

Mutations. Morphologically demonstrable chromosome alterations and functional alterations of genetic material. Mutations may be classified as follows:

I. Gene mutations—hereditary alterations resulting in new alleles.

II. Chromosomal mutations occurring as morphological change:

1. Numerical chromosome changes
   (a) change of diploid chromosome number (aneuploidy);
   (b) reduplication or twofold reduction of complete haploid sets (polyploidy, haploidy).
2. Structural chromosome changes
   (a) intrachromosomal rearrangements (deletions, duplications, changes of ordered gene arrangement (inversions);
   (b) interchromosomal rearrangements altering gene group linkages (translocations, transpositions).

N

Nucleosome. Cornberg and Thomas identified, in 1974, an octamer of histones (H2A), (H2B), (H3), (H4); this reiterative chromatin unit, defined as nucleosome, contains DNA of about 200 base pairs in length.

P

Paris Conference (1972), another update on the human karyotype nomenclature with respect to improvements in chromosome banding techniques, karyotype identification and aberration characterization.

Polyploidy. Endomitotic increase in a haploid set by an even number. Chromosomes do not separate since their centromeres are unable to go apart.

Progeria. A hereditary disease which affects the repair system. Two variant forms of it are progeria of children (the Hutchinson-Gilford syndrome) and progeria of adults (Werner's syndrome). The Hutchinson-Gilford syndrome is a rare genetic condition. Already in early postnatal months infants show symptoms of aging, loss of subcutaneous fat, alopecia and severe atherosclerosis affecting cerebral and cardiac function. Adolescent patients look gerontal and die before the age of 20.

R

Repair of DNA. Recovery of the native structure of damaged DNA at the molecular level.

Replication of DNA. Production of an exact copy of the two-stranded DNA molecule by gene duplication during cell division.

Ribonucleic acid (RNA). A polymer made up by ribose-containing nucleotides (AMP, CMP, GMP, UMP). RNA has key functions in ribosomal protein synthesis.

Ribosomes. RNA-protein particles the function of which in protein synthesis is to attach amino acid residues into sequences determined by messenger RNA. Several ribosomes bound to mRNA make a polyribosome, or a polysome.

S

Satellite DNA. DNA centrifugation in cesium chloride density gradient yields a fraction which is phyknographically demonstrable, apart from the main chromosomal DNA peak, as a satellite peak designating the satellite DNA.

Stalk. An acentric constriction (an unwound region of a wound chromosomal strand) which often contributes to the nucleolus formation. An interchangeable term for it is nucleolar constriction.

Sex chromatin, or Barr body. Occurs on the nuclear membrane or within the nucleus in females. Absent in males.

T

Transcription. Production of an RNA strand complementary to one of DNA molecule strands.

Translation. Protein production in the ribosome. Genetic information carried in messenger RNA nucleotide sequences is decoded to determine polypeptide chain amino acid sequences.

Transposition. Insertion into a chromosome region of a small fragment with genes that are unusual for the region, i.e. are brought into it from another genomic site. The smallest transposable fragments (500 to 1500 nucleotide pairs) are known to occur in bacteria and termed insertional segments. Larger fragments (3000 to 25 000 base pairs) are called transposons, or transposable elements.

U

Unscheduled DNA synthesis. A relatively low synthesis induced by irradiation in non-S-phase cells.

X

X chromosome. A chromosome associated with sex determination that is present in two copies in the homogametic sex and in one copy in the heterogametic sex.

Xeroderma pigmentosum. An inherited defect in the repair of ultraviolet-light damage DNA, associated with extreme sensitivity to sunlight and multiple skin cancers.

Y

Y chromosome. The sex chromosome that is present only in the heterogametic sex; in mammals, the male-determining sex chromosome.

Z

Zinc finger. A structural motif found in many DNA-binding proteins, in which a fingerlike projection entraps a zinc ion.

Zygote. The product of the fusion of a female and a male gamete in sexual reproduction; a fertilized egg.

REFERENCES

Abbo G., Zellweger H., Cuany R. Satellite association (SA) in familial mozaicism. *Helv. Paediatr. Acta,* 1966, 21, 4, 293-299.

Abruzzo M., Mayer M., Jacobs P. Aging and aneuploidy: evidence for the preferential involvement of the inactive X-chromosome. *Cytogenet. and Cell Genet.,* 1985, 39,4,275-278.

Akesson H., ForrsmanH. A Study of Maternal Age in Down's Syndrome.*Ann.Hum Genet.,* 1966, 29, 3, 271-276.

Akesson H., Wahlstrom J. The length of the Y chromosome in men examined by forensic psychiatrists. *Hum. Genet.,* 1977, 39, 1-5.

Akifiev A., Grishanin A. Some biological aspects of chromatin diminution. *J. Obshchei Biol.,* 1993, 54, 1, 5-16. (In Russian)

Akifiev A., Grishanin A., Degtiaryv S. Chromatin diminution is a key process explaining the eukaryotic genome size paradox and some mechanisms of genetic isolation. *Genetika.* 2002, 38, 5,595-606.

Akman S., O'Connor., Robriguez H. Mapping oxidative DNA damage and mechanisms of repair. *Ann NY Acad. Sci.,* 2000, 899, 88-102.

Alhadeff B., Cohen M. Frequency and distribution of sister chromatid exchanges in human peripheral lymphocytes. *Isr. J. Med. Sci.* 1976, 12, 1440-1447.

Allen F., Latt S. In vivo BrdR-33258 Hoechst analysis of DNA replication kinetics and sister chromatid exchange formation in mouse somatic and meiotic cells. *Chromosoma,* 1976, 58, 325-340.

Allison D., Nestor A. Evidence for a relatively random array of human chromosomes on the mitotic ring. *J Cell Biol ,* 1999. 145, 1-14.

Almagon M., Cole R. Changes in chromatin structure during the aging of cell cultures as revealed by differential scanning calorimetry. *Biochemistry* 1989, 28, 13, 5688-5693.

Almagon M., Cole R. Differential scanning calorimetry of nuclei as a test for the effects of anticancer drugs on human chromatin. *Cancer. Res.* 1989b, 49, 20, 5561-5566.

Amchenkova A., Narovlyansky A., Stonova N., et al. Stability of C-band heterochromatin polymorphism in two human stable cell lines differing in their susceptibility to Coxsackie B3 virus. *Cytologia i Genetika,* 1981, 15, 2, 31-34. (In Russian)

Andres L. *Introduction to Human Kariology.* M.L. Medgiz, 1934.

Andriadze M., Pleskach N., Mikhelson V., Zhestianikov V. Spontaneous and induced sister chromatid exchanges in blood lymphocytes of donors and patients with Xeroderma Pigmentosum under the action of inhibitors of DNA repair and replication (novobiocin, 3-metoxybenzamide and caffeine). *Cytologia*, 1986, 28, 1, 69-85. (In Russian)

Andrulis, E.D., Neiman A.M., Zappulla D.C., Sternglanz R., Perinuclear localization of chromatin facilitates transcriptional silencing. *Nature,* 1998, 394, 592-595.

Antoshchina M., Poryadkova N. Technique for differential staining of sister chromatids without using fluorochromes. *Cytologia i Genetika*, 1978, 12, 4, 349-352. (In Russian)

Aragatzuni K. Association of acrocentric chromosomes in 80-85-year-old persons. *Genetika*, 1967, 6, 79-85. (In Russian)

Arce M. The effect of donor sex and age on the number of sister chromatid exchanges in human lymphocytes growing in vitro. *Hum. Genet.,* 1981, 1, 83-85.

Archeu T., Cole H., Cordin gley M., et. al. Differential steroid hormone induction of transcription from the mouse mammary tumoy virus premotor. *Mol. Endocrinol.* 1994, 8, 568-576.

Ardito G., Lambert L., Bigatti P., Stanyon R. The effect of cell kinetics time on SCE and NOR associations in Macaca fucata lymphocytes. *Cytogenet. and Cell Genet.,* 1983, 36, 3, 532-536.

Ardito G., Lamberti L., Brogger A. Satellite associations of human acrocentric chromosomes identified by trypsin treatment at metaphase. *Ann. Hum. Genet.,* 1978, 41, 4, 455-462.

Arrighi F., Hsu T. Localization of heterochromatin in human chromosomes. *Cytogenetics,* 1971, 10, 81-86.

Asadov Sh., Berdyshev G. Familial longevity in the Azerbaijan SSR. *J. Clin. and Exp. Gerontol.,* 1986, 8, 1-2, 75-88.

Auerbach A.,Greenbaum J.,Pujara K., et al. Spectrum of sequence variation in the FANCG gene: an International Fanconi Anemia Registry (IFAR) study. *Hum. Mutat.,* 2003, 21, 2, 158-168.

Avivi L., Feldman M., Bushuk W. The mechanism of somatic association in common wheat, Triticum Aestivum L. suppression of somatic association by colchicine. *Genetics,* 1969, 62, 745-752.

Bablishvili N. *Variability of heterochromatin induced by an inorganic salt of sodium depending on age.* Author's abstract (of dissertation), Tbilisi, 2002.

Bailey S., Brenneman M., Goodwin E. Frequent recombination in telomeric DNA may extend the proliferative life of telomerase-negative cells. *Nucleic Acids Res.* 2004, 32, 12, 3743-3751.

Bain A., Gauld I. Chromosome endoreduplication in cultures of necropsy spleen. *Lancet,* 1964, 1, 7339, 936 337.

Bairamyan T. Sister chromatid exchanges in human lymphocytes as observed using 5-bromodeoxyuridine. *Genetika*, 1978, 14, 6, 1085-1092. (In Russian)

Bairamyan T., Zakharov A. Sister chromatid exchanges under bromine incorporation into cytosine nucleotides of DNA. I. Combined using of 5-bromodeoxycytidine and thymidine as a method of bromine incorporation into cytosine. *Cytologia* , 1978, 20, 5, 507-513.

Bajnoczky K., Mehes K. Age-dependence of centromere separation sequence. *Clin. Genet.,* 1985, 28, 5, 412-412. (In Russian)

Bajnoczky K., Mehes K. Parental centromere separation sequence and aneuploidy in the offspring. *Hum. Genet.,* 1988, 78, 3, 286-288.

Baker W. Mechanisms of chromosomal and gene inactivation in Drosophila. Proceedings of the thirteenth International Congress of Genetics. Part 1. Symposium 6. Gene and Chromosome inactivation. *Genetics,* 1974, 78, 1, 33- 41.

Balicek P., Zizka J., Skalska H. Variability and familial transmission of constitutive heterochromatin of human chromosomes evaluated by the method of linear measurement. *Hum. Genet.,* 1978, 42, 3, 257-265.

Barenfeld L. Down's syndrome: pathogenesis, radiorezistant DNA synthesis and chromosomal instability. *Cytologia,* 2002, 44, 4, 379-386.

Barja G. Endogenous oxidative stress: relationship to aging, Longevity and caloric restriction. *Ageing Res. Rev.,* 2002, 1, 3, 397-411.

Barr M., Bertram E. A morphological distinction between neurons of the male and female and the behaviour of the nucleolar satellite during accelerated nucleoprotein synthesis. *Nature,* 1949, 163, 4148, 676-677.

Barton D., David F., Fix E. Random points in circle and the analysis of chromosome patterns. *Biometrika,* 1963, 50, 1-2, 23-29.

Barton D., David F., Merrington M. The relative positions of the chromosomes in the human cell in mitosis. *Ann. Hum. Genet.,* 1965, 29, 139-146.

Bartram C.,Koske-Westphal T.,Passarge E. Chromatid exchanges in ataxia telangiectasia, Bloom syndrome, Werner syndrome and Xeroderma pigmentosum. *Hum. Genet.,* 1976, 40, 1, 79-86

Battaglia A., Loprieno N. Micronucleated lymphocytes in people occupationally exposed to potential environmental contaminants: the age effect. *Mut. Res.,* 1991, 256, 1, 13-20.

Becak M., Becak W., Pereira A. Somatic pairing, endomitosis and chromosome aberrations in snakes (Viperidae and Colubridae). *Ann. Acad. Bras. Cienc.,* 2003, 75, 3, 285-300.

Beckman K., Ames B. The free radical theory of aging matures. *Physiol. Rev.,* 1998, 77, 547-581.

Bednar F., Horowitz R., Grigoriev S., Carruthers L., Hansen Z., Koster A., Wood cock. C. Nucleosomes, Linkey DNA, and linkey histone form a uniqui structural motif that directs the highey-order folding and compaction of chromatin. *Cell Biol.,* 1998, 95, 24, 14173-14178.

Beek B., Obe G. The human leucocyte test system VI. The use of sister chromatid exchanges as possible indicators for mutagenic activities. *Humangenetik,* 1975, 29, 2, 127-134.

Beerman W. Nuclear differentiation and functional morphology of chromosomes. Cold Spring Harbor. *Symp. Quant. Biol.,* 1956, 21, 171-217.

Bell A. *The duration of life and conditions associated with longevity. A study of the Hyde Genealogy.* Genealogical Records Office, Washington, 1918.

Belloni M., Cozri R., De Marco G., Venezia O. Distribuzione delle aberrazioni in cellule somatiche di gangli di Drosophila melanogaster irradiate in interfase (G_2) in mitosi. *Atti. Assoc. Genet. Ital.,* 1977, 22, 133-134.

Beltran J., Forbes W., Robertson B., Page B. Human Y-chromosome variation in normal and abnormal babies and their fathers. *Ann. Hum. Genet.,* 1979, 42, 3, 315-325.

Bender M., Griggs H., Bedford J. Mechanisms of chromosomal aberration production III. Chemicals and ionizing radiations. *Mutat. Res.* 1974, 23, 2, 197-212.

Bender M., Preston R., Leonard R. et al. Chromosomal aberration and sister-chromatid exchange frequencies in peripheral blood lymphocytes of a large human population sample. 11. Extension of age range. *Mutat. Res.,* 1989, 212, 2, 149-154.

Berlowitz L. Chromosomal inactivation and reactivation in mealy bug. Proceedings of the Thirteenth International Congress of Genetics. Part 1. Symposium 6. Gene and chromosome inactivation. *Genetics,* 1974, 78, 1, 311-322.

Birshteyn Ya. Heterochromatin, the mosaic effect of gene localization in Drosophila and a problem of chromosome heterochromatinization. *Adv. Modern Biol.*, 1976, 81, 2, 225-243.(In Russian)

Bishun N. Association of human satellite chromosomes. *Experientia,* 1966, 22, 223-224.

Bjorkstein J. *Cross linkage and the aging process in the theoretical aspects of aging.* (Ed. Rockstain M.) Academic Press, New York, 1977, 43-59.

Bjorkstein J., Tenhu H. The cross linking theory of aging - added evidence. *Exp. Gerontol.,* 1990, 25, 2, 91-95.

Bloom A., Archer P., Awa A. Variation in the human chromosome number. *Nature,* 1967, 216, 487-489.

Bloom A., Hell F., Tsuchimoto T., Weilinger K. Chromosomal breakage in leukocytes of South American Indians. *Cytogenet. Cell Genet.,* 1973, 12, 175-186.

Bloom S., Goodpasture C. An improved technique for selective silver staining of nucleolar organizer regions in human chromosomes. *Hum. Genet.,* 1976, 34, 199-206.

Bochkov N. *Human chromosomes and irradiation.* M., Atomizdat, 1971.

Bochkov N., Kozlov V., Pilosov R., Sevankaev A. The level of spontaneous chromosome aberrations in human leukocyte cultures. *Genetika,* 1968, 4, 6, 94-98. (In Russian)

Bochkov N., Kuleshov N. Dependence of the chemical mutagenesis intensity in human cells on sex and age. *Genetika,* 1971, 7, 3, 132-138. (In Russian)

Bochkov W., Kuleshov N., Zhurkov V. Analysis of spontaneous chromosome aberrations in human leucocyte cultures. *Cytologia,* 1972, 14, 10, 1267-1278. (In Russian)

Bohr V., Evans M., Fornace A. Biology of Disease. DNA repair and its pathogenetic implication. *Lab. Invest.,* 1989, 61, 2, 143-161.

Bolshev L. and Smirnov N. *Tables of mathematical statistics,* M., 1968.

Bond S., Singh S. Genotype-specific reduction in methyl nitrosourea (MNU)-induced sister chromatid exchanges (SCE) in vivo during aging, *Experientia,* 1988, 44, 9, 782-785.

Bouchlaka C., Abdelhak S., Amouri A. Fanconi anemia in Tunisia: hige prevalence of group A and identification of new FANCA mutations. *J. Hum. Genet.,* 2003, 48, 7, 352-361.

Bownian P., Meek R., Daniel C. Decreased syntheses of nucleolar RNA in aging human cells in vitro. *Exp. Cell Res.,* 1976, 101, 434-437.

Brewen J., Peacock W. The effect of tritiated thymidine on sister chromatid exchange in a ring chromosome. *Mutat. Res.,* 1969, 7, 3, 433-440.

Brodsky V., Khavinson V.,Zolotarev Y., et al. The rhythm of protein synthesis in hepatocytes of rats of different age. The norm and the effect of peptide Livagen *.Izv. Akad. Nauk Ser.Biol.,* 2001, 5, 517-521.

Brogger A. Apparently spontaneous chromosome damage in human leukocytes and the nature of chromatid gaps. *Humangenetik,* 1971, 13, 1-14.

Brogger A. Is the chromatid gap a folding defect due to protein change? Evidence from mercaptoethanol treatment of human lymphocyte chromosomes. *Hereditas,* 1975, 80, 131-136.

Bross K., Krone W. Ribosomal cistrons and acrocentric chromosomes in man. *Humangenetik,* 1973, 18, 71-75.

Brown D., Gurdon J. Absence of ribosomal RNA synthesis in sister chromatid exchanges in Bloom's syndrome lymphocytes. *Proc. Nat. Acad. Sci., USA,* 1974, 71, 11, 4508-4512.

Brown S. Heterochromatin. Heterochromatin provides a visible guide to suppression of gene action during development and evolution. *Science,* 1966, 151, 3709, 417-425.

Brown W., Epstein I., Little I. Progeria cells are stimulated to repair DNA by co-cultivation with normal cells. *Exp. Cell Res.,* 1976, 97, 2, 291-296.

Buck S., Sandmeier F., Smith F. RNA polymerase I propagates andirechoval spreading of rDNA selent chromatin. *Cell,* 2002, III, 7, 1003-1014.

Buckton K., Evans G. Methods of analysis of human chromosome aberration. *JOH, Geneva,* 1975, 7-64.

Buckton K., O'Riordan M., Jacobs P., et al. Q- and G-band polymorphisms in the chromosomes of three human populations. *Ann. Hum. Genet.,* 1976, 40, 99-112.

Budjakov B., Zolotarjov B. Problem of arrangement of chromosomes in human somatic cell nuclei. *Adv. Modern Genet.,* 1971, 43, 254-295.

Bukvic N., Gentile M., Susca F., et al. Sex chromosome loss, micronuclei, sister chromatid exchange and aging: a study including 16 centenarians. *Mutat. Res.,* 2001, 498, 1-2, 159-167.

Burnet M. *Intrinsic mutagenesis: A genetic approach to aging.* New York. Wiley, 1974.

Buys C., Osinga J., Anders G. Age dependent variability of ribosomal RNA-gene activity in man as determined from frequencies of silver stain nucleolar organizing regions on metaphase chromosomes of lymphocytes and fibroblasts. *Mech. Aging and Dev.,* 1979, 11, 55-75.

Cadotte M., Fraser D. Etude de l'aneuploidie observee dans les cultures de sang et de moelle en fonction du membre et de la longue des chromosomes de chaque groupe et de l'age de deux sexes des subjects. *Union Med. Can.,* 1970, 99, 11, 2003-2007.

Calow P. Bidders hypothesis revised solution to some key problems associated with general molecular theory of aging. *Geronthol.,* 1978, 24, 448-458.

Campisi J. Aging, chromatin and food restriction-connecting the dots. *Science,* 2000, 289, 2062-2063.

Campisi J. From cells to organisms: can we learn about aging from cells in culture? *Exp. Gerontol.,* 2001, 36 ,4-6, 607-618.

Cannavo G., Paiardini M., Galati P., Cervasi B., Montroni M., De vico G., Guetard J., Bocchi-no M., Picerno T., Magnani M., Silvestri G. Piedimonte G. Abnormal intracellular kinetics of Cell-Cycle-dependent proteins in lymphocytes from patients infected with human immunodeficiency virus: a novel biologic link between immune activation accelerated T cell turnover, and high lenels of apoptosis. *Blood,* 2001, 97, 6, 1756-1764.

Capoa A., Ferraro M., Menender F., et al. Age staining of the nucleoulus organizer (NO) and its relationship to satellite association. *Hum. Genet.,* 1978, 44, 71-77.

Capoa A., Rocchi A., Giogliani F. Frequency of satellite association in individuals with structural abnormalities of nucleolar organizer region. *Humangenetik,* 1973, 18, 111-115.

Cardellini E.,Cinelli S.,Gianfranceschi G., et al. Differential scanning calorimetry of chromatin at different levels of condensation. *Mol. Biol. Rep.,* 2000, 27, 3, 175-180.

Carnevale A., Ibafier B., del Castillo V. The segregation of C- band polymorphisms on chromosome 1, 9 and 16. *Amer. J. Hum. Genet.,* 1976, 28, 412-416.

Carrano A., Wolff Sh. Distribution of sister chromatid exchanges in the euchromatin and heterochromatin of the Indian muntsack. *Chromosoma,* 1975, 53, 4, 366-369.

Carvalho C., Pereiza H., Ferreira J., et al. Chromosomal G-dark bands determine the spetial organization of centromeric heterochromatin in the nucleus. *Mol. Biol. Cell,* 2001, 12, 3563-3572.

Catalan J., Autio K., Wessman M., et al. Age-associated micronuclei containing centromeres and the X-chromosomes in lymphocytes of women. *Cytogenet. Cell Genet.,* 1995, 68, 1-2, 11-16.

Catania J., Fairweather D. DNA methylation and cellular aging. *Mutat. Res.,* 1991, 256, 283-293.

Cavazza B.,Bizzolara G., Lazzarini G., et al. Thermodynamics of condensation of nuclear chromatin. A differential scanning calorimetry of the salt dependent structural transitions. *Biochemistry,* 1991, 30, 37, 9060-9070.

Chadov B. Spontaneous arising of translocations and non-homologues pairing. *Genetika,* 1975, 11, 1, 91-101. (In Russian)

Chadov B. and Chadova E. Non-homologous pairing and spontaneous interchanges between non-homologous chromosomes in Drosophila melanogaster males. *Genetika,* 1977, 13, 3, 477-481. (In Russian)

Chaganti R., Schonberg S., German J. A manifold increase in sister chromatid exchanges in Bloom's syndrome lymphocytes. *Proc. Nat. Acad. Sci.,* USA, 1974, 71, 11, 4508-4512.

Chandra H., Nyrayana A., Buche V., Hungerford D. Peripheral location of the human late and homologous association of autosomes numbers one, two and three. *J. Genet,* 1972, 61, 78-83.

Chebotarev A. Quantitative analysis of sister chromatid exchanges in cell. *Genetika,* 1979, 15, 8, 1392-1398. (In Russian)

Chemitiganti S., Verma R., Ved Brat S., Dosik H. Random single chromatid type segregation of human acrocentric chromosomes in BrdU-labeled mitoses. *Can. J. Genet. and Cytol.,* 1984, 26, 2, 137-140.

Chicago Conference: Standardization in human Cytogenetics. *Orig. Art. Ser.,* 1966, 2, 3-9.

Chieds B. Prospects for genetic screening. *J. Pediat.,* 1975, 87, 1125-1132.

Chirkhirzhina E., Vorobev V. Linker histones: conformational chances and role in the structural organization of chromatin. *Cytologia,* 2002, 44, 8, 721-736. (In Russian)

Claire P., Lewis G., Marc T., et al. Programmed cell death of the normal human noutrophilan in vitro model of senescens. *Microsc. Res. and Techn.,* 1994, 28, 4, 327-344.

Clarkson J., Evans H. Unscheduled DNA synthesis in human leucocytes after exposure to UV-light, γ-rays and chemical mutagens. *Mutat. Res.,* 1972, 14, 413-430.

Clarkson J., Peiter R. Repair of X-ray damage in aging WI-38 cells. *Mutat. Res.,* 1974, 23, 107-112.

Claussen U., Michel S., Muhlig P et al. Demistifying chromosome preparation and the implications for the concept of chromosome condensation during mitosis. *Cytogenet. Genome Res.,* 2002, 98, 136-146.

Clayson D., Mehta R., Iverson F. International commission for protection against environmental mutagens and carcinogens oxidative DNA damage - the effects of certain genotoxic and operationally non-genotoxic carcinogens. *Mutat. Res.,* 1994, 317, 1, 25-42.

Cleaver J. Repair processes for photochemical damage in mammalian cells. *Adv. Pediat. Res.,* 1974, 4, 1-75.

Cohen M., Shaw M. The association of acrocentric chromosomes in 1000 normal human male metaphase cells. *Ann. Hum. Genet.,* 1967, 31, 129-137.

Comings D. The rationale for an ordered arrangement of chromatin in the interphase nucleus. *Amer. J. Hum. Genet.,* 1968, 20, 440-460.

Comings D. Arrangement of chromatin in the nucleus. *Hum. Genet.* 1980, 53, 131-143.

Conger A. Real chromatide deletions versus gaps. *Mutat. Res.,* 1967, 4, 449-459.

Cooke P. On the inheritance of differentiated traits, *Biol. Rev.,* 1971, 49, 51- 84.

Cooke P., Curtis D. General and specific patterns of acrocentric association in parents of Mongol children. *Humangenetik,* 1974, 23, 279-287.

Cooper E., Barkhan P., Hale A. Mitogenic activity of Phytohemagglutinin. *Lancet,* 1961, 2, 7195, 210-210.

Cornforth M., Eberle R. Termini of human chromosomes display elevated rates of mitotic recombination. *Mutagenesis,* 2001, 16, 85-89.

Cortes –Gutierres E., Cerda-Flores R.,Silva-Cudish J et al., Evaluation of sex chromosome aneuploidies in women with Turner's syndrome G-banding and FISH. A serial case study. *J. Reprod. Med.* 2003,48,10,804-808.

Costa M. Molecular targets of nicel and chromium in human and experimental systems. *Scand. J. Work Environ. Health.,* 1993, 19, 71-74.

Cottliar A., Fundia A., Moran C., et. al. Evidence of chromosome instability panereatitis. F. *Exp. Clin Cancer Res.* 2000, 19, 4, 513-517.

Counter C., Avilion A., Le Feuvre C., et al. Telomere shortening associated with chromosome instability is arrested in immortal cells express telomerase activity. *EMBO Journal,* 1992, 11, 5, 1921-1929.

Court Brown W., Buckton K., Jacobs P., et al. Chromosome studies on adults. In.: *Eugen Lab., Mem. XIII,* 1966, 1-31.

Craig-Holmes A., Shaw M. Polymorphism of human constitutive heterochromatin. *Science,* 1971, 174, 702 -704.

Craig-Holmes A., Moore F., Shaw M. Polymorphism of human C-band heterochromatin. II. Family studies with suggestive evidence for somatic crossingover. *Amer. J. Hum. Genet.,* 1975, 27, 2, 178-189.

Crastes de Paulet A. Radicaux libres et vieillissement. *Ann. Biol. Clin.,* 1990, 48, 5, 323-330.

Cremer T., Kreth G., Koester H., et al. Chromosome territorye interchromatin domain compartment, and nuclear matrix an integrated view of the functional nuclear architecture. *Crit. Rev. Eucaryot Gene Expr.* 2000, 12, 179-212.

Cristofalo V., Pignolo R. Replicative senescence of human fibroblast - like cells in culture. *Physiol. Rev.,* 1993, 73, 617-638.

Croft J., Bridger J., Boyle S., et al. Differences in the localization and morphology of chromosomes in the human nucleus. *J Cell Biol.,* 1999, 145, 6, 1119-31.

Crossen P. DNA replication and sister chromatid exchange in 9hg+. *Cytogenet. and Cell Genet.,* 1983, 35, 2, 152-155.

Crossen P., Drets M.., Arrighi F., Gohnston D. Analysis of the frequency and distribution of sister chromatid exchanges in cultured human lymphocytes. *Hum. Genet.,* 1977, 35, 3, 345-352.

Cutler R. Evolution of human congenity and the genetic complexity governing aging rate. *Proc. Nat. Acad. Sci. USA,* 1975, 72, 4664-4668.

Cutler R. Alteration with age in the informational stage and flow systems of the mammalian cell-Brief Defects. *Original Article Series,* 1978, 14, 1, 463-498.

D'Anna J., Tobey R., Barham S., Gurley L. A reduction in the degree of H4 acetylation during mitosis in Chinese hamster cells. *Biochem. Biophis. Res. Commun.,* 1977, 77, 1, 187-194.

Das C., Kaufmann B., Gay H. Histone-protein transition in Drosophila melanogaster. *Exp. Cell Res.,* 1964, 35, 3, 507-514.

De Boer J., Andressoo J.,de Wit., et al. Aging in mice deficient in DNA repair and transcription. *Science,* 2002, 296, 1276-1279.

Deknudt G., Lionard A. Aging and radiosensitivity of human somatic chromosomes. *Exp. Gerontol.,* 1977, 12, 5-6, 237-240.

De la Chapell A., Schroder J., Selander R. Repetitious DNA in mammalian chromosomes. Hereditas, 1971, 69, 149-153.

De La Jerna T., Carlson K., Imbalzano A. Mammalian JWI/SNF complexes promote MyoD-mediated muscle differentiation. *Nat. Genet.,* 2001, 27, 187-190.

Demoise C., Conard R. Effects of age and radiation exposure on chromosomes in a Marshall Island population. *J. Gerontol.,* 1972, 27, 2, 197-201.

Denton T., Liem S., Cheng K., Barrett J. The relationship between aging and ribosomal gene activity in human as evidenced by silver staining. *Mech. Aging and Dev.,* 1981, 15, 1-7.

Denver Conference. A proposed standard system of nomenclature of human mitotic chromosomes. *Lancet,* 1960, 1, 7133, 1063-1065.

Dernburg, A.F., Broman K.W., Fung J.C., et al., Perturbation ofnuclear architecture by long-distance chromosome interactions. *Cell,* 1996, 85, 745-759

Deryagin G., Iordansky A. Phenotypic variability of satellited chromosomes. *Genetika,* 1971, 7, 10, 13-22. (In Russian)

Dewey W., Noel J., Dettor C. Changes in radiosensitivity and dispersion of chromatin during the cell cycle of syncronous Chinese hamster cells. *Rad, Res.,* 1972, 52, 373-394.

Dittes H., Krone W., Bross K., et al. Biochemical and cytogenetic studies on the nucleolus organizing regions (NOR) of man. II. A family with the 15/21 translocation. *Humangenetik,* 1974, 26, 47-59.

Drings P., Sonnemann E. Euchromatisierung menschlicher lymphocyten durch phytohamagglutinin. *Res. Exp. Med.,* 1974, 164, 63-76.

Dubinin N.,Tarasov B. Molecular mechanisms of chromosome aberrations formation. *Adv.Modern.Genet.,* 1979,8,113-146.

Dubrovsky I., Berdyshev G. Changes in the compactions of the liver chromatin nucleosomes in cattle during aging. *Ukrain.Biokh. J.,* 186, 58, 3, 8-13. (In Russian)

Dumont M., Bello M., Guichaoua M., Luciani J. Differential associative behaviour of mitotic and meiotic acrocentric chromosomes. *Hum. Genet.,* 1989, 82, 1, 35-39.

Durante M,. George K., Wu H., Yang T. Rejoining and misrejoining of radiation-induced chromatin breaks. 1. Experiments with human lymphocytes. *Radiat. Res.,* 1996, 145, 3, 274-280.

Dvalishvili N. Study of the spontaneous level of quantitative and structural changes of chromosomes at in vitro aging. *Proc. Acad. Sci. Georgian SSR,* 1983, 112, 1, 140-144. (In Russian)

Dvalishvili N. *Characteristic of human chromosomes in long-term cultures of lymphocytes (in vitro) aging).* Author's abstract (of dissertation), Tbilisi,1989. (In Russian)

Egolina N., Davydov A., Benyush V., Zakharov A. Genetic determination of nucleolar organizer activity in human chromosomes. *Bull. Exp. Biol. and Med.,* 1981, 91, 3, 350-363. (In Russian)

Egolina N., Zakharov A. Spiralization of Chinese hamsters chromosomes after 5-bromodeoxyuridine treatment of cells in two mitotic cycles. *Cytologia,* 1972, 14, 2, 165-171. (In Russian)

Ellis N., Groden J., Straughen J., et al. The Bloom's syndrome gene product homologous to rec helicases. *Cell,* 1995, 88, 655-666.

Ellis J., Penrose L. Enlarged satellites and multiple malformations in the same pedigree. *Ann. Hum. Genet.,* 1960, 25, 159-162.

El Zawahry M., Kishin A. Chromosome changes in relation to human aging. *Eur. J. Cell Biol.,* 1980, 22, 1, 553-557.

Engman F. Satellite count on mitotic chromosomes of monozygotic and dizygotic human twins. *Lancet,* 1967, 2, 7526, 1114-1116.

Engman F. Morphology of human mitotic chromosomes. *Adv. Teratol.,* 1972, 5, 13-50.

Ercal N., Guzer-Orhan H., Aykin-Burns N. Toxic metals and oxidative stress part I: mechanisims involved in metal-induced oxidative damage. *Curr. Top. Med. Chem.,* 2001, 1, 6, 529-539.

Esposito D., Fassina G.,Szabo P., et al. Chromosomes of older humans are more prone to aminopterine-induced breakage. *Proc.Nat. Acad. Sci. USA,* 1989, 86, 4, 1302-1306.

Evans H. Chromosome aberration and target theory. In: *Radiation-induced Chromosome Aberrations.* (Ed. S. Wolff), Columbia Univ. Press. New York, 1963, 8-40.

Evans H., Buckland R., Pardue M. Location of the genes coding for 18S and 28S ribosomal RNA in the human genome. *Chromosoma,* 1974, 48, 405-426.

Evans H., Swery O. *The chromosomes in man. Sex and somatic.* Univ. of California, 1929.

Evans M., Bohr V. Gene-specific DNA repair of UV-induced cyclobutane pyrimidine dimers in some cancer-prone and premature aging human syndromes. *Mutat. Res.,* 1994, 314, 221-231.

Evans R., Norman A. Unscheduled incorporation of thymidine in ultraviolet-irradiated human lymphocytes. *Rad. Res.,* 1968, 36, 287-298.

Fainaru O., Lichtenberg D., Pinchuk I., et al. Preelclampsia is association with increased susceptibility of serum lipids to cooper-induced peroxidation in vitro. *Acta Obstet. Gynecol . Scand.,* 2003, 82, 8, 711-715.

Fang J., Jagillo G., Ducayen M., Graffeo J. Aging and X-chromosome loss in the human ovary. *Obstet. Gynec.,* 1975, 454, 455-458.

Fedak G., Helgason S. Somatic Association of Chromosomes in Barley. *Can. J. Genet and Cytol.,* 1970, 12, 496 500.

Feldman M., Mello-Sampayo T., Sears E. Somatic Association in Triticum Aestivum. Proc. *Natl. Acad. Sci. USA,* 1966, 56, 1192-1196.

Fenech M., Dreosti I., Rinaldi J. Folate, vitamin B12, homocysteine status and chromosome damage rate in lymphocytes of older men. *Carcinogenesis,* 1997, 18,7, 1329-1336

Ferguso-Smith M. The sites of nucleolus formation in human pachytene chromosomes. *Cytogenetics,* 1964, 3, 12, 124-134.

Ferguson-Smith M., Handmaker S. Observation on the satellite human chromosomes. *Lancet,* 1961, 1, 638 640.

Fernandez R., Barragan M., Bullejos M., et al., New C-band protocol by heat denaturation in the presence of formamide. *Hereditas,* 2002, 137, 145-148.

Fesenko E., Morozov K. About the dependence of different sensitivity of cellular ages and time after irradiating. In: *Cellular mechanisms of genetic processes (mutagenesis, reparation),* M., Nauka, 1976, 192 199.

Fischer P., Golob E., Kunse-Muhl E., et al. Chromosomal aberrations in peripheral blood cells in man following chronic irradiation from internal deposits of thorotrast. *Radiat. Res.,* 1966, 29, 505-509.

Fischer P., Kim M. Technique for the visualization of exchange aberrations in human chromosomes. *Exp. Pathol.,* 1975, 10, 34, 216-219.

Fitzgerald P. A mechanism of X-chromosome aneuploidy in lymphocytes of aging women. *Humangenetik,* 1975, 28, 153-158.

Ford F., Woolam D. Significance of variation in satellite incidence in normal human mitotic chromosomes. *Lancet,* 1967, 2, 7505, 26-27.

Ford J., Laster P. Fractions affecting the displacement of human chromosomes from the metaphases plate. *Cytogenet. and Cell Genet.,* 1982, 33, 4, 327-332.

Forne A. Reference values for chromosome aberration in human lymphocytes as indicators of genetoxic effects. *Sci. Total. Environ.,* 1992, 120, 1-2, 149-153.

Forne A. Benzene-induced chromosome aberrations: A follow-up study. *Environ. Health Perspectives* (Supp), 1996, 1309-1312.

Fourel G., Lebrun E., Gilson E. Protosilencers as bulding bloks for heterochromatin. *Bioessays,* 2002, 24, 9, 828-835.

Fransz P, Soppe W, Schubert I. Heterochromatin in interphase nuclei of Arabidopsis thaliana. *Chromosome Res.,* 2003, 11, 3, 227-240.

Freguet B. Protein repair and degradation during aging. *Scientific World Journal,* 2002, 2, 1, 248-254.

Froland A., Mikkelsen M. Studies on satellite associations in human cells. *Hereditas,* 1964, 52, 2, 248-248.

Frolkis V. Self-regulation processes and mechanisms of aging. *Izvestia, Acad. Med. Nauk SSSR,* 1986, 10, 8-15. (In Russian)

Frolov A. The frequency of acrocentric chromosome associations in the long-term cultures of human lymphocytes. *Cytologia i Genetika,* 1986, 20, 3, 166-171. (In Russian)

Furey T., Haussler D. Integration of the cytogenetic map with the draft human genome sequence. *Hum. Mol. Genet.,* 2003, 12, 9,1037-1044.

Galloway S., Evans. H. Sister chromatid exchange in human chromosomes from normal individuals and patients with ataxia telangiectasia. *Cytogenet. and Cell Genet.,* 1975, 15, 1, 17-29.

Galperin-Lemaitre H., Hens L., Kirsch-Volders M., Susanne Ch. Non-random association of trypsin-banded human acrocentric chromosomes. *Hum. Genet.,* 1977, 35, 3, 261-268.

Galperin-Lemaitre H., Hens L., Sele B. Comparison of acrocentric associations in male and female cells. Relationship to the active nucleolar organizers. *Hum. Genet.,* 1980, 54, 349-353.

Ganguli B. Cell division, chromosomal damage and micronucleus formation in peripheral lymphocytes of healthy donors: related to donor's age. *Mut. Res.,* 1993, 295, 3, 135-148.

Gartenberg M.R. The Sir proteins of Saccharomyces cerevisiae: mediators of transcriptional silencing and much more.*Curr.Opin.Microbiol. 2000,* 3, 132-137.

Gaubatz J., Prashad N., Cutler R. Ribosomal RNA gene-dosage as a function of tissue and age for mouse and human. *Hum. Biochem. Biphys. Acta.,* 1976, 418, 3, 358-376.

Geigl J., Langer S., Barwisch S., Pfleghaar K., Lederer G., Speicher M. Analysis of gene expression patterns and chromosomal changes associated with aging. *Cancer Res.* 2004, 64, 8550-8557.

Geitler L. Die Enistehung der Polyploiden Somakerne der Heteropteren durch Chromosomenteilung ohne Kernteilung. *Chromosoma,* 1939, 1, 1, 1-22.

German J. Synthesis of deoxyribonucleic acid during interphase. *Lancet,* 1962, 1, 7232, 744-744.

German J. Cytological evidence for crossingover in vitro in human lymphoid cells. *Science,* 1964, 144, 298 301.

Ghosh S., Talukder G., Sharma A. Frequency of chromosome aberrations induced by trimethylchloride in human peripheral blood lymphocytes in vitro: relation to age of donors. *Mech. Aging and Dev.,* 1991, 57, 2, 125-137.

Ghosh S., Talukder G., Sharma A. Chromosomal alterations and sister chromatid exchanges induced by zirconium oxychloride in human lymphocytes in vitro with relation to age of donors. *Mech Aging and Dev.,* 1992, 62, 3, 245-254.

Gibson D., Prescott D. Induction of sister chromatid exchanges in chromosomes of rat Kangaroo cells by tritium incorporated into DNA. Exp. *Cell Res.,* 1972, 74, 2, 397-402.

Gigliani F., De Capoa A., Rocchi A. A marker chromosome number 14 with double satellite observed in two generations and unbalanced chromosome constitution associated with normal phenotype. *Humangenetik,* 1972, 15, 191-195.

Gille J. Vrije radicalehen veroudering. *Pharm. Weekbl.,* 1990, 125, 21, 523-528.

Gindilis V. The spiralization of mitotic chromosomes and karyogramic analysis in men. *Cytologia,* 1966, 8, 2, 144-157. (In Russian)

Giulotto E., Mottura A., De Carli L., Nuzzo F. DNA repair in UV-irradiated heteroploid cells at different phases of the cell cycle. *Exp. Cell Res.,* 1978, 113, 2, 415-420.

Gladyshev G. Thermodynamic theory of aging. Exploins reasons of aging and death with standpoint of the general laws of nature. *Adv. Gerontol.,* 2001, 7, 42-45.

Goh-Kongoo O. Sister chromatid exchange in the aging population. *J.Med.,*1981, 12, 2-3, 195-198.

Goldstein S. The role of DNA repair in aging of cultured fibroblasts from Xeroderma pigmentosum and normals. *Proc. Soc. Exp., Med.* 1971, 137, 730-734.

Goldstein S. Aging in Vitro. Growth of Cultured Cells from the Galapagos Tortoise. *Exp. Cell Res.,* 1974, 83, 297-302.

Goldstein S. Replicative senescence: The human fibroblast comes of age. *Science,* 1990, 249, 1129-1133.

Goldstein S., Srivastava A., Riabewel K., Shmookler R. Genetic organization and expression in aging human fibroblasts. *In vitro,* 1985, 21, 3, Pt 2, 14-14.

Goletz T., Smith J., Pereira-Smith O. Molecular genetics approaches to the study of cellular senescence. *Cold Spring Harb. Symp. Quant. Biol.,* 1994 59, 59-66.

Goodman R., Fechheimer N., Miller F., et al. Chromosomal alterations in three age groups of human females. *Amer. J. Med. Sci.,* 1969, 258, 26-34.

Goodpasture C., Bloom S., Hsu T., Arrighi F. Human nucleolus organizers: the satellites of the stalks? *Amer. J. Hum. Genet.,* 1976, 28, 559-566.

Gopal A., Arora N., Vardanian S., Messineo F. Utility of transesophageal echocardiography for the characterization of cardiovascular anomalies associated with Terner's syndrome. *J. Am. Soc. Echocardiogr.,* 2001, 14, 1, 60-62.

Gosden J., Gosden C., Lawrie S., Buckton R. Satellite DNA loss and nucleolar organizer activity in an individual, with a de novo chromosome 13, 14 translocation. *Clin. Genet.,* 1979, 15, 518-529.

Goya R. Role of programmed cell death in the aging process: an unexplored possibility. *Gerontology,* 1986, 32, 1, 37-42.

Graham G., Garcia M., Cummings M. A tandemly repeated, nonribosomal DNA sequence found on human chromosomes that engage in satellite associations. *Amer. J. Hum. Genet.,* 1984, 36, 4, Suppl. 1-138.

Greaves I., Rens W., Ferguson-Smith M., et al. Conservation of chromosome arrangement and position of the X in mammalian sperm suggests functional significance. *Chromosome Res.,* 2003, 11,5,503-512.

Greider C. Telomeres. Current opinion. *Cell Biol.,* 1991, 3, 3, 444-451.

Grell R. Pairing at the chromosomal level. *J. Cell physiol.* 1967, 70, suppl., 1-119.

Grell R., Day J. Chromosome pairing in the oogonial cells of Drosophila melanogaster. *Chromosoma,* 1970, 31, 434-445.

Griffin D. The incidence, origin and etiology of aneuploidy. *Int. Rev. Cytol.,* 1996, 167, 263-296.

Grouchy I., Thiefery S., Arthuis M., et al. Chromosomes marqueurs familiaux et aneuploidie. Role possible de l'interaction chromosomique. *Ann. Genet.,* 1964, 7, 76 83.

Grumbach M., Morishima A., Chu E. Of the sex chromatin and the sex chromosomes in sexual anomalies in man; relation to origin of the sex chromatin. *Acta Endocrinol.,* 1960, 35, suppl. 31, 633.

Gruneberg H. The case for somatic crossingover in the mouse. *Genet. Res. Camb.,* 1966, 7, 1, 58-75.

Guanti C., Petrinelli P. DNA and acrocentric chromosomes in man. *Cell Different.,* 1974, 2, 314-324.

Gusev V. Free radical theory of ageing in the gerontological paradism. *Adv. Gerontol.,* 2000, 4, 41-49.

Guttenbach M., Schakowski R., Schmid M. Aneuploidy and aging: sex chromosome exclusion into micronuclei. *Hum. Genet.,* 1994, 94, 3, 295-298.

Guttenbach M., Koschorz B., Bernthaler U., et al. Sex chromosome loss and aging in situ hybridization studies on human interphase nuclei. *Amer. J. Hum. Genet.,* 1995, 57, 5, 1143-1150.

Haaf T., Schmid M. Experimental condensation inhibition in constitutive and facultative heterochromatin of mammalian chromosome. *Cytogen. Cell Genet.,* 2000, 91, 113-123.

Hahn H. The regulation of protein synthesis in the aging cell. *Exp. Gerontol.,* 1970, 5, 323-334.

Hahn H. Failures of regulation mechanisms as causes of cellular aging. *Adv. Gerontol. Res.,* 1971, 3, 1-38.

Hahn H., Miller J., Eichhorn G. Age-related alterations in the structure of nucleoprotein. IV. Changes in the composition of whole histone in rat liver. *Gerontology,* 1969, 15, 4-5, 293-301.

Hamerton J., Taylor A., Angell R., M. Guire V. Chromosome investigations of a small isolated human population: chromosome abnormalities and distribution of chromosome counts according to age and sex among the population of Tristan da Cunha. *Nature,* 1965, 206, 1232-1234.

Hando J., Nath J., Tucker J. Sex chromosomes micronuclei and aging in women. *Chromosoma,* 1994, 103, 3, 186-192.

Hansen J. Conformational dynamics of the chromatin fiber in solution: determinants, m. *Annu Rev Biophys Biomol Struct.,* 2002, 31 , 361-392

Hansson A. Satellite association in human metaphases. A comparative study of normal individuals, patients with Down syndrome and their parents. *Hereditas,* 1979, 90, 1, 59-83.

Hara R., Mo F., Samcay A. DNA Damage in the Nucleosome Core is Refractory to Repair by Human Excision Nuclease. *Mol. Biol.* 2000, 20, 24, 9173-9181.

Harbers E., Sandritter W. Aufbau und funktion des chromatins. *Klin. Wschr.,* 1973, 51, 631-643.

Hardner K., Hogstedt B., Kolnig A., et al. Sister chromatid exchanges and structural chromosome aberrations in relation to age and sex. *Hum. Genet.,* 1982, 62, 4, 305-309.

Harman D. Aging: A theory based on free radicals and radiation chemistry. *J. Gerontol.,* 1956, 11, 298-300.

Harman D. Free radicals in aging. *Mol. and Cell Biochem.,* 1988, 84, 2, 155-161.

Harman D. Aging: Minimizing free radical damage. *J. Anti Aging Medicine,* 1999, 2, 15 – 36.

Harris C., Connor R., Jajson F., Leberman M. Intracellular distribution of DNA repair synthesis induced by chemical carcinogens or ultraviolet light in human diploid fibroblasts. *Cancer Res.,* 1974, 34, 12, 3461-3468.

Harrison D. Normal function of transplanted marrow cell lines from aged mice. *J. Gerentol.,* 1975, 30, 279-285.

Hart R., D'Ambrosio S., Ng. K., Modak S. Longevity, stability and DNA repair. *Mech. Aging and Dev.,* 1979, 9, 3-4, 203-223.

Hart R., Setlow R. DNA repair in late-passage human cells. *Mech. Aging. and Dev.,* 1976, 5, 67-77.

Hassold T., Chiu D. Maternal age-specific rates of numerical chromosome abnormalities with special reference to trisomy. *Hum. Genet.,* 1985, 70, 1, 11-17.

Hasty P., Vijg J. Genomic priorities in ageing. *Science,* 2002, 296, 1250-1251.

Hawley R., Arbel T. Yeast genetics and the fall of classical view of meiosis. *Cell,* 1993, 72,301-303.

Hayflick L. The limited in vitro lifetime of human diploid cells strains. *Exp. Cell Res.,* 1965, 37, 614-636.

Hayflick L. The biology of human aging. *Amer. J. Med. Sci.* 1973, 265, 432-445.

Hayflick L. Cell senescence and cell differentiation in vitro. In: *Aging and development* (Ed. H. Bredt and M.W. Rohen). Stuttgart. Schattauer Verlag, 1975, 4, 1-15.

Hayflick L. Cell biology of aging. *Fed., Proc.,* 1979, 38, 5, 1847-1850.

Hayflick L. Intracellular determinants of cell aging. *Mech. Aging and Dev.,* 1984, 28, 2-3, 177-185.

Hayflick L. The future of aging. *Nature,* 2000, 408, 6809, 267-269.

Hayflick L. Theories of biological aging. *Exp. Gerontol.* 1985, 20, 3-4, 145-159.

Hayflick L., Moorhead P. The serial cultivation of human diploid cell strains. *Exp. Cell Res.,* 1961, 25, 4, 585-621.

Heitz E. Heterochromatin, chromocentren, chromomeren. *Ber. Dtsch. Bot. Ges.,* 1929, 47, 274-284.

Heits E. Die Ursache der gestrambigen Zahl. Lage, Form und Grösse der pflanzlichen Nucleolen. *Planta,* 1931, 12, 4, 775-844.

Heliot L., Mongelard F., Klin C., O'Donohue M., Chassery F., Robert-Nicoud M., Usson Y. Nonrandom Distribution of Metaphase AgNOR steining Patterns on Human Acrocentric Chrorosomes. F. Histochem. *Cytochem.,* 2000, 48, 13-20.

Henderson A., Mosckowits G., Warburton D. Do numerical polymorphisms exist at the human 5s locus? *Hum. Genet.,* 1980, 54, 1, 83-85.

Henderson A., Warburton D., Atwood K. Location of rDNA in the human chromosome complement. *Proc. Natl. Acad. Sci. USA,* 1972, 68, 3394-3398.

Heneen W., Nikhols W. Nonrandom arrangement of metaphase chromosomes in cultured cells of the Indian deer Muntiacus muntjak. *Cytogenetics,* 1972, 11, 153-164.

Hens L., Kirsch-Volders M., Arrighi F., Susanne C. Relationship between measured chromosome distribution parameters and Ag-staining of the nucleolus organizer regions. *Hum. Genet.,* 1980, 53, 3, 363-370.

Hens L., Kirsch-Volders M., Susanne C., Galperin-Lemaitre H. Relative position of trypsin-handed homologous chromosomes in human metaphase figures. *Humangenetik,* 1975, 28, 3, 303-311.

Hensler P., Annab L., Barrett J., Pereira-Smith O. A gene involved in control of human cellular senescence on human chromosome 1q. *Mol. Cell Biol.,* 1994, 14, 4, 2291-2297.

Herrera F., West K., Schiltz R., Nakatani Y., Bustin M. Histone H1 is a Specific Repressor of core histone Acetylation in Chromatin Mol. and *Cell. Biol.,* 2000, 20, 2, 523-529.

Heslop-Harrison J., Bennet M. Chromosome order-possible implications for development. *J. Embryol. and Exp. Morphol.,* 1984, 85, Suppl., 51-73.

Hilliker A., Appels R. The arrangement of interphase chromosomes: structural and functional aspects. *Exp. Cells Res.,* 1989, 185, 2, 297-318.

Hiraoka Y., Agard D., Sedat J. Temporal and spatial coordination of chromosomal movement, spindle formation, and nuclear envelope breakdown during prometaphase in Drosophila melanogaster embryos. *J Cell Biol.,* 1990, 111: 2815-2828.

Hoehn H., Au K., Karp L., Martin G. Somatic stability of variant C-band heterochromatin. *Hum. Genet.,* 1977, 35, 163-168.

Hoehn H., Martin G. Non-random arrangement of human chromatin: topography of disomic markers X, Y and Ih. *Cytogenet. and Cell Genet.,* 1973, 12, 443-452.

Hoehn H., Nagel M., Kranze W. In vitro alteration of human acrocentric chromosomes. *Humangenetik,* 1971, 11, 146-154.

Hoffman S., Hromas R., Amemiya C., Mohrenweiser H. The location of MZF-1 at the telomere of human chromosome 19q makes it vulnerable to degeneration in aging cells, *Leuk. Res.,* 1996, 20, 3, 281-283.

Holmquist G., Comings D. Sister chromatid exchanges and chromosome organization based on bromodeoxyuridine gemsa-C-banding technique. *Chromosoma,* 1975, 52, 3, 245-259.

Holms R., Keating M., Cork A., et al. Loss of the Y-chromosome in acute myelogenous leukemia: report of 13 patients. Cancer Genet. Cytogenet., 1985, 17, 3, 269-278.

Hoo J., Perslow I. Relation between the SCE points and the DNA replication bands .*Chromosoma,* 1979, 71, 3, 67-74.

Hornsby P. Cellular senescence and tissue aging in vivo. J. Gerontol., 2002, 57, 251-256.

Howard B. Replicative senescence: considerations relating to the stability of heterochromatin domains. *Exper. Gerontol.,* 1996, 31, ½, 281-293.

Hsu T. A possible function of constitutive heterochromatin: The bodyguard hypothesis. *Genetics,* 1975, 79, Suppl., 137-150.

Hsu T., Moorhead P. Chromosome anomalies in human neoplasma with special reference to the mechanisms of polyploidization and aneuploidization in the HeLa strain. *Ann. N.Y. Acad. Sc.,* 1956, 63, 6, 1083-1094.

Hsu T., Pathak S. Differential rates of sister chromatid exchanges between euchromatin and heterochromatin. *Chromosoma,* 1976, 58, 3, 269-273.

Hulbert A. Life, death and membrane bilayers. *J. Exp. Biol.,* 2003, 206, 14, 2303-2311.

Ikushima T., Wolff Sh. Sister chromatid exchanges induced by light flashes to 5-bromodeoxyuridine and 5-iododeoxyuridine substituted Chinese hamster chromosomes. *Exp. Cell Res.,* 1974, 87, 1, 15-19.

Intano G., McMahan C., McCarrt J., et al. Base excision repair is limited by different proteins in male germ cell nuclear extracts prepared from young and old mice. *Mol. Cell Biol.,* 2003, 22, 7, 2410-2418.

Iritani B., Delrow J., Grandori C., et al. Modulation of T-lymphocyte development, growth and cell size by the Myc antagonist and transcriptional repressor Mad 1. *EMBO Journal,* 2002, 21, 18, 4820-4830.

Izakovic V., Vahancik A. Caryotyp Lymfocytov obvodovej krivi a buniek kostnej drone u 80-rochych a starsich aosob. *Vnitrini Lek.,* 1984, 30, 10, 974-983.

Jacobs P., Brunton M., Court Brown W., et al. Change of human chromosome count distributions with age: evidence for a sex difference. *Nature,* 1963, 197, 1080-1081.

Jacobs P., Court Brown W., Doll R. Distribution of human chromosome counts in relation to age. *Nature,* 1961, 191, 1178-1180.

Jarvik L., Falek A., Kallmann F., Lorge I. Survival trends in a senescent twin population. *Amer. J. Hum. Genet.,* 1960, 12, 170-179.

Jarvik L., Fu-Sun Yen, Moralishvili M. Chromosome examinations in aging institutionalized women. *J. Gerontol.,* 1974, 29, 3, 269-276.

Jarvik L., Fu-Sun Yen, Tsu-Ker Fu, Matsuyama S. Chromosomes in old age: a six year longitudinal study. *Hum. Genet.,* 1976, 33, 17-22.

Jarvik L., Kato T. Chromosome examinations in aged twins. *Amer. J. Hum. Genet.,* 1970, 22, 562-573.

Joachimiak A. Uklad chromosomow w jadrze interfazowym. *Post. Biol. Komorki,* 1987, 14, 3, 237-254.

John B., Miklos G. Functional aspects of satellite DNA and heterochromatin. *Rev. Cytol.,* 1979, 58, 1-114.

Johnson E., Umbenhauer D., Hill R., et al. Karyotypic and phenotypic changes during in vitro aging of human endothelial cells. *J. Cell Physiol.,* 1992, 150, 17-27.

Johnson T. A personal retrospective on the genetics of aging. *Biogerontology,* 2002, 3, 1-2, 7-12.

Jokhadze T. Effect of cadmium on spontaneous mutation process in Crepis capillaris seeds. In: *Theor-Pract.Probl.Com. Mol. Genet., M., Nauka,* 1977, 48-51.

Jokhadze T., Lezhava T. Study of chromosome structural changes induced by heavy metal salts at in vivo and in vitro aging. *Genetika,* 1994, 30, 12, 1630-1632.(In Russian)

Kadotani T., Ohama K., Nakayama T., et al. Chromosome aberrations in leukocytes of normal human adults from 49 couples. *Proc. Japan. Acad.* 1971, 47, 724-728.

Kadotani T., Watanabe Y., Kadotani T. Chromosome study in aging. *Int. J. Human Genet.* 2002, 2, 1, 5-9.

Kadotani T., Watanabe Y., Makine S. The incidence of satellite associations in D- and G-group chromosomes and maternal aging. *Proc. Japan Acad.,* 1978, 54, ser. B, 6, 277-282.

Kallmann F. Twin data on the genetics of aging. Aba Found. *Cell Aging,* 1975, 3, 131-148.

Kang Y., Koo D., Park J., et al. Differential inheritance modes of DNA methylation between euchromatic and heterochromatic DNA sequences in ageing bovine fibroblasts. *FEBS Lett.,* 2001, 498, 1, 1-5.

Kanungo M. Alterations in gene expression during senescence. *Proc. Indian Acad. Sci.,* 1984, 93, 173-177.

Kanungo M. *Biochemistry of aging.* M., Mir. 1982. (In Russian)

Kappeler L., Epelbaum J. Biological aspects of Longevity and ageing. *Rev. Epidemiol Sante Publique.* 2005, 53, 3, 235-241.

Karn J., Johnson E., Vidali G., Allfrey V. Differential phosphorylation and turnover of acidic proteins during the cell cycle of synchronized HeLa cells. *J. Biol. Chem.,* 1974, 249, 3, 667-677.

Kato H. Induction of sister chromatid exchanges by chemical mutagens and its possible relevance to DNA repair. *Exp. Cell Res.,* 1974a, 85, 2, 239-247.

Kato H. Spontaneous sister chromatid exchanges detected by a BUdR-labelling method. *Nature,* 1974b, 6, 251, 5470, 70-72.

Kato H., Coetzee M., Ove P. A comparison of DNA repair synthesis in primary hepatocytes from young and old rats. *Mech. Aging and Dev.,* 1985, 29, 3, 283-298.

Kato H., Stich H. Sister chromatid exchanges in ageing and repair deficient human fibroblasts. *Nature,* 1976, 260, 447-448.

Kawanishi S., Hiraku Y., Murata M., Oikawa S. The role of methals in site-specific DNA damage with reference to carcinogenesis. Free Radic. Biol. Med., 2002, 32, 9, 822-832.

Kellum R. Is HP1 an RNA detector that functions both in repression and activation? *J. Cell Biol.,* 2003, 161, 4, 671-672.

Kennah H., Coetzee M., Ove P. A comparison of DNA repair synthesis in primary hepatocytes from young and old rats. *Mech. Ageing and Dev.,* 1985, 29, 3, 283-288.

Kerkis Yu., Rajabli S. The caryotype changes in human somatic cells related to age.*Cytologia,*1966,8,2,282-285.

Kerkis Yu., Rajabli S., Pospelova T., Visotskaya L. On the cause of progressing with age aneuploidy in human leukocytes. *Genetika,* 1967, 4, 137-141. (In Russian)

Khavinson V. Kh. Peptides and Ageing. *Neuroendocrinol. Lett.,* 2002, 23 (suppl. 3),3-144.

Khavinson V., Lezhava T., Monaselidze J., et al. Effect of Livagen on chromatin in lymphocytes from old people. *Bull Exp. Biol. Med.,* 2002, 134,4, 389-392.

Khavinson V., Lezhava T., Monaselidze J., et al. Peptide Epitalon activates chromatin at the old age. *Neuroendocrinology letters,* 2003, 24, 5, 329-333.

Khesin R., Leibovitch B. Chromosome structure, histones and gene activity in Drosophila. *Mol. Biol,* 1976, 10, 1, 3-33. (In Russian)

Khochbin, S., Verdel, A., Lemercier, C., Seigneurin-Berny. Functional significance of histone deacetylase diversity. *Curr Opin Genet Dev.,* 2001, 11, 2, 162-166.

King C., Gillespie E., McKenna P., Barnett Y. An investigation of mutation as a function of age in humans. *Mutat. Res.,* 1994, 316, 2, 79-90.

King M., Wilson A. Evolution at two levels in humans and chimpanzees. *Science,* 1975, 188, 107-116.

Kipshidze N., Todria M., Djavakhishvili N. Changes in blood system in longevity. In: *Problems of hematology and blood transfusion,* 1964, 10, 2, 32-36.

Kirkwood T. DNA, mutations and aging. *Mutat. Res. DNA Aging: Gen. Instab. and Aging.*, 1988, 219, 1, 1-7.

Kirsch-Volders M., Hens L., Ashley T., et al. Comparison of Chromosome association in human lymphocytes arrested by colcemid, nodazole or cycloheximide. *Can. J. Genet. and Cytol.* 1983, 25, 3, 304-311.

Kirsch-Volders M., Hens L., Verschaeve L., Driesen M., Susanne C. Chromosome distribution as a parameter for mutagenesis, *Mutat. Res.*, 1978, 53, 2, 210-210.

Kirsch-Volders M., Weltens R., Hens L., Susanne C., Defrise-Gussenhoven E. Variability of NOR-staining in monozygotic and dizygotic twins *Eur. J. Cell Biol.*, 1980, 12, 1, 117-120.

Kitani Y. Orientation arrangement and association of somatic chromosomes. *Japan. J. Genet.*, 1963, 38, 244 256.

Kleisner de Galan E. Age and Chromosomes. *Nature*, 1966, 211, 1324-1325.

Kong Z. Analysis of mitomicin induced sister chromatid exchanges. *Acta Zool. Sin.*, 1988, 34, 3, 210-213.

Konoplya E., Detinkin O., Zhitkovich A. Study of age-related differences in the structural state of rat liver cell chromatin using micrococcal nuclease and circular dichroism. *Biokhimia*, 1988, 53, 11, 1876 1882. (In Russian)

Koronberg J., Freedlender E. Giemsa technique for the defection of sister chromatid exchanged. *Chromosoma*, 1974, 48, 4, 355-360.

Kowarzyk H., Sternhaus H., Szymaniec S. Arrangement of chromosomes in human cells. IV. Distances between centromeres in the metaphase disc. *Bull. Acad. Polon. Sci. Biol.*, 1967, 15, 21-25.

Kram D., Scheneider E., Tice R., Gianas P. Aging and sister chromatid exchanges. *Exp. Cell. Res.*, 1978, 114, 2, 471-475.

Krzanowska H., Bilinska B. Number of chromocentres in the nuclei of mouse Sertoli cells in relation to the strein and age of males from puberty to senescence. *J. Reprod. Fertil.*, 2000, 118, 2, 343-350.

Kubai D. Nonrandom chromosome orange arrangements in germ-like nuclear of Sciara coprophila males: the basis for nonrandom chromosome segregation on the meiosis I spindle. *J. Cell Biol.*, 1987, 6, 2433-2446.

Kucherenko N., Tsudzevich B., Blouma Ya., Babeniuk Yu. *Biochemical model of chromatin activity regulation. Kiev, Naukova Dumka*, 1983, 3-244. (In Russian)

Kuleshov N. Age sensitivity of human chromosomes to cytosinearabinoside applied at phase G_2 of the cell cycle. *Cytologia*, 1972a, 16, 11, 1368-1372. (In Russian)

Kuleshov N. Chromosome aberrations in cultured leukocytes of individuals of different sex and age induced by degranol. *Genetika*, 1972b, 7, 3, 123-131. (In Russian)

Kurnit D. Satellite DNA and heterochromatin variants: the case for unequal mitotic crossingover. *Hum. Genet.*, 1979, 47, 169-186.

Kuznetsov A., Krushalov A., Ilyushenko V., Zycova I., Kudritsky Yu. Age and sex dependence on spontaneous frequency of chromosome aberrations in human peripheral blood lymphocytes. 1. Age dynamics of spontaneous frequency of chromosome aberrations and aberrant cells in peripheral blood lymphocytes. *Genetika*, 1980, 16, 7, 1285-1230. (In Russian)

Kuznetsova S., Zaritskaya M. The polymorphism of heterochromatin sites in chromosomes 1, 9, 16 and Y observed in long-living individuals and in persons of different age from two regions of the Soviet Union. *Cytologia i Genetika,* 1986, 20, 6, 409-416. (In Russian)

Kuznetsova S, Zaritskaia M.Chromosomes. Aging. Longevity. *Tsitol Genet.* 1986, 20, 4, 304-313.

Kyng K., Bohr V. Gene expression and DNA repair in progeroid syndromes and human aging. *Ageing Res. Rev.* 2005, 4, 4, 579-602.

Lamb M.,*Biology of aging.* Blackie Glasgow and London, 1977.

Lambert B., Hansson K., Lindsten J., et al. Bromodeoxyuridine-induced sister chromatid exchanges in human lymphocytes. *Hereditas,* 1976, 83, 2, 163-174.

Lambert B., Ringborg U. Decreased DNA repair synthesis in higher ages. *Mut. Res.,* 1976, 38, 2, 129-130.

Lambert B., Ringborg U., Swanbek G. Repair of UV-induced DNA lesion in peripheral lymphocytes from healthy subjects of various ages, individuals with Down's syndrome and patients with actinic keratosis. II Workshop: 13. *Mut. Res.,* 1977, 46, 133-134.

Lange K., Page B., Elston R. Age trends in human chiasma frequencies and recombination fractions. I. Chiasma frequencies. *Amer. J. Hum. Genet.,* 1975, 27, 3, 410-418.

Langst G., Beckey B. Nucleeosome mobilization and positioning by FSWJ-containing chromatin-remodeling factors. F. Cell Science., 2001, 114, 2561-2568.

Lansdorp P. Role of telomerase in hematopoietic stem cells. *Ann. N.Y. Acad Sci.* 2005. 1044, 220-227.

Latt S. Localization of sister chromatid exchanges in human chromosomes. *Science,* 1974, 185, 4145, 74-76.

Latt S., Stetten G., Fuergens L., et al. Induction by alkylating agents of sister chromatid exchanges and chromatid breaks in Fancont's anemia. Proc. *Nat. Acad. Sci. USA,* 1975, 72, 10, 4066-4070.

Lee C., Klopp R., Weinarnen R., Prolla T. Gene expression profile of aging and its retardation Coloric Restraction. *Science,* 1999, 285, 1390-1393.

Lee H., Habas R., Abate-Shen C., MSX1 cooperates with histone H1b for inhibition of transcription and miogenesis. *Science*, 2004, 304, 1675-1678.

Lernid R., Riva M., Delpra L., Ginelli E. Satellite associations and silver staining in a case of multiple G and D variants. *Hum. Genet.,* 1980, 53, 2, 237-240.

Levan A., Hauschka T. Endomitotic reduplication mechanisms in ascites tumors of the mouse. *J. Nat. Cancer Inst.,* 1953, 14, 1, 1-43.

Levy M,. Allsopp R., Futcher A., Greider C., Harley C. Telomere end-replication problem and cell aging. *J. Mol. Biol.,* 1992, 225, 4, 951-960.

Lewis E. Adv. Genet., 1950, 3, 73. In: Birstein Ya. Heterochromatin, the mosaic effect of gene localization in Drosophila and a problem of chromosome heterochromatinization. *Adv. Modern Biol.,* 1976, 81, 2, 225-243. (In Russian)

Lewis M. PRELP, collagen and a theory of hutchinson-Gilford progeria. Aging Res. Rev., 2003, 2, 1, 95-105.

Lezhava T. Morphology of human acrocentric chromosomes. *Cytologia,* 1966, 8, 2, 286-290. (In Russian)

Lezhava T. Endoreduplication in the Stein-Leventhal syndrome. *Cytologia,* 1968, 10, 2, 241-246. (In Russian)

Lezhava T. Analysis of the centromere heterochromatin of persons aged from 80 to 114. *Proc. Acad. Sci. Georgian SSR*, 1977, 3, 5, 465-473. (In Russian)

Lezhava T. Association of human chromosomes at senile age. *Cytologia i Genetika,* 1979, 13, 481-485. (In Russian)

Lezhava T. Heterochromatinization - one of the leading factors of aging. *Cytologia i Genetika,* 1980, 14, 3, 71-76. (In Russian)

Lezhava T. The activity of nucleolar organizer regions of human chromosomes in extreme old age. *Gerontology,* 1984a, 30, 94-99.

Lezhava T. Heterochromatinization as a key factor in aging. *Mech. Ageing and Dev.,* 1984b, 28, (2-3), 279-288.

Lezhava T. Sister chromatid exchange in human lymphocytes in extreme age. *Proc. Japan Acad.,* 1987, 63, Ser. B, 369-372.

Lezhava T. *Human chromosomes in very senile age,* State University, Tbilisi, 1991, 3-257. (In Russian)

Lezhava T. Role of Heterochromatinization of human Chromosomes in Aging. *Ind. J., Hum. Genet.,* 1996, 2: 33-42.

Lezhava T. *Chromosome in very senile age: 80 years and over.* M.,"Nauka", 1999.

Lezhava T. Chromosome and aging: genetic conception of aging. *Biogerontology,* 2001a, 2, 4, 253-260.

Lezhava T. Human chromosome functional characteristics and aging . *Adv. Gerontol.,* 2001b, 8, 34-43.

Lezhava T., Chitashvili R., Khmaladze E. Use of the mathematical "satellite model" for associations of acrocentric chromosomes depending on human age. *Bio-Medical Computing,* 1972, 3, 101-199.

Lezhava T. and Khmaladze B. Spontaneous level of quantitative structural changes of chromosomes in the senile age. *Proc. Acad. Sci. Georgian SSR,* 1978, 4, 2, 162-170. (In Russian)

Lezhava T., Chitashvili R. Variability of homologous chromosomes location in metaphases of human somatic cells depending on age. *Proc. Acad. Sci. Georgian SSR,* 1979, 5, 6, 547-553. (In Russian)

Lezhava T., Gogniashvili O., Gvazava E. Characteristics of the functional organization of chromatin in old age. *Bull. Acad. Sci. Georgian SSR,* 1979, 96 1, 157-160. (In Russian)

Lezhava T., Prokofieva V., Mikhelson W. Reduction of UV-induced unscheduled DNA synthesis in human lymphocytes at the extreme old age. *Cytologia,* 1979a, 21, 1360-1363. (In Russian)

Lezhava T., Chitashvili R. Sister chromatid exchanges in human lymphocytes in extreme old age. *Cytologia,* 1982, 24, 59-65. (In Russian)

Lezhava T., Khmaladze E. Aneuploidy in human lymphocytes in extreme old age. *Proc. Japan Acad.,* 1988a, 64, Ser. B, 128-130.

Lezhava T,. Khmaladze E. Characteristics of Cis- and Transorientation chromatid types of association in human extreme old age. *Proc. Japan Acad.,* 1988b, 64, Ser. B, 131-134.

Lezhava T., Khmaladze E. Characterization of Cis- and Transposition of chromatids of the association type at human extreme old age. *Genetika,* 1989, 15, 4, 727-733. (In Russian)

Lezhava T., Dvalishvili N. Cytogenetic and biochemical studies on the nucleolus organizing regions of chromosomes in vivo and in vitro aging. *Age,* 1992, 15, 41-43.

Lezhava T., Monaselidze J., Chanchalashvili Z., et al. Study of condensed euchromatin level in extreme old age by differential scanning calorimetry. *Bull. Acad. Sci. of Georgia,* 1993, 19, 334-337.

Lezhava T., Bablishvili N. Reactivation of heterochromatin induced by sodium hydrophospate at the old age. *Proc. Georg. Acad. Sci. Biol,* Ser.B, 2003, 1, 1-2, 1-5.

Lezhava T., Khavinson V.,Monaselidze J., et al. Bioregulator Vion-induced Reactivation of Chromatin in Cultured Lymphocytes from old People. *Biogerontology,* 2004, 4, 73-79.

Liem S., Denton T., Cheng K. Distribution patterns of satellite associations in human lymphocytes relative to age and sex. *Clin. Genet.,* 1977, 12, 2, 104-110.

Lima de Faria A., Birnsteil M., Faworska U. Amplification of ribosomal cistrons in the heterochromatin of Acheta. *Genetics,* 1969, 61, supp. 1, 145-159.

Lin M., Alfi O. Detection of sister chromatid exchanges G-U-diamidino-2-phenylindole fluorescence. *Chromosoma,* 1976, 57, 3, 219-225.

Lindsey J., McGill N., Lindsey L., et al. In vivo loss of telometric repeats with age in humans. *Mutat. Res.,* 1991, 256, 1, 45-58.

Lints A. *Genetics and aging.* 14 Series. Editor H.P. von Hahn. Basel, 1978.

Little J. Relationship between DNA repair capacity and cellular aging. *Gerontology,* 1976, 22, 28-55.

Little J., Epstein J., Williams J. DNA repair and survival of Y-irradiated human diploid cells. *In vitro,* 1974, 9, 3-4, 350.

Lobov I. and Podgornaya O. The role of the nuclear matrix proteins in heterochromatin assembly. *Cytlogia,* 1999, 41, 562-573. (In Russian)

London Conference. The normal human karyotype. *Cytogenetics,* 1963, 2, 4-5, 254-268.

Luckinbil L., Foley P. Experimental and empirical approaches in the study of aging. *Biogerontology,* 2000, 1, 3-13.

Lundgren M.,Chow C.,Sabbattini P. et al. Transcription factor dosage affects changes in higher order chromatin structure associated with activation of heterochromatic gene. *Cell,* 2000, 103,733-743.

Lyon M. Sex chromatin and gene action in the mammalian X-chromosome. *Amer. J. Hum. Genet.,* 1962, 14, 2, 135-167.

Machavariani M. Cytogenetic effects of X-rays in the primary culture of Macaca Mulatta kidney cells. *Cytologia i Genetika,* 1976, 10, 1, 12-14. (In Russian)

Maeng S., Chung H., Kim., Lee B., Shin Y., Kim S., Yn I. Chromosome aberration and lipid peroxidation in chromium-exposed workers. *Biomarkers.,* 2004, 9, 6, 418-434.

Mamaev N., Mamaeva S. Structure and function of nuclear organizer region of chromosome. Molecular,cytological and clinical aspects.*Cytölogia,* 1992,34, 3-24.

Mantovani M., dos Santos Abell L.,Mestriner C.,et al. Accentuated plymorphism of heterochromatin and nucleolar organizer regions in Astyanax scabripnnis (Pisces, Charocidae) tools for understanding kariotypic evolution. *Genetica,* 2000, 109, 3, 161-168.

Manuelidis L., Borden J. Reproducible compartmentalization of individual chromosome domains in human CNS cell revealed by in situ hybridization and three dimensional reconstruction. *Chromosoma* (Berl.), 1988, 96, 397-410.

Maramatsu S.,Hanada H., Himeno K. Effects of cadmium in biological tissues by atomic absorption induction in bonne marrow cells and spermatogonia on mice. *Radiobiological Equivalents. Chem. Pllit. Advis Group Meet.,* 1980.

Marcus M., Sperling K. Condensatin-inhibition by 33258 Hoechst of centromeric heterochromatin in prematurely condensed mouse chromosome. *Exp., Gell., Res.,* 1979, 123, 406-411.

Marin G., Prescott D. The frequency of sister chromatid exchanges following exposure to varying dots of H^3-thymidine or X-rays. *Cell Biol.,* 1964, 21, 159-167.

Martin G. Genetic and environmental modulations of chromosomal stability: their roles in aging and oncogenesis. *Ann. N.Y. Acad. Sci.,* 1991, 621, 401-417.

Martin G. Genetic and evolutionary aspects of aging. *Fed. Proc.,* 1979, 38, 6, 1962-1967.

Martin G., Kellett M., Kahn G. Aneuploidy in cultured human lymphocytes 1. Age and sex difference. *Age and Aging,* 1980, 9, 3, 147-153.

Martin G.,Oshima J. Lessons from human progeroid syndroma. *Nature,* 2000, 408, 263-266.

Martin G., Spraque C., Epstein C. Replicative life-span of cultivated human cells-effects of donor's age. *Lab. Invest.,* 1970, 23, 86-92.

Martin H., Leutert G., Rother P., Rotzsch W. Hypothesen des Alterns-Stand und Probleme. Z. *Alternsforsch.,*1986, 2, 67-77.

Martin M.,Genesca A., Latre L., Ribas M., et al. Radiation-induced chromosome breaks in Ataxia-telangiectasia cells remain open. *Int. J. Radiat. Biol.,* 2003, 79, 3, 203-210.

Martin R., Rademaker A. The effect of age on the frequency of sperm chromosomal abnormalities in normal men. *Amer. J. Hum. Genet.,* 1987, 41, 3, 484-492.

Mattern R., Cerutti P. Age dependence of excision of ray-damaged thymine by isolated nuclei from diploid human lung fibroblasts WI-38. *Nature,* 1975, 254, 450-452

Mattevi M., Salzano F. Effect of sex, age and cultivation time on number of satellite and acrocentric associations in man. *Humangenetik,* 1975b, 29, 265-270.

Mattevi M., Salzano F. Senescence and human chromosome changes. *Humangenetik,* 1975a, 27, 1-8.

Mazin A. Enzymatic DNA metilation as an aging mechanism. *Molekulyarnaya Biologia.* 1994,28,1,21-51.

McClintock B. The production of homozygous deficient tissues with mutant characteristics by means of the aberrant mitotic behavior of ring-shaped chromosomes. *Genetics,* 1938, 23, 315-376.

Mc Dowell T., Gibbons R., Sutherland H., et al. Localization of putative transcriptional regulator (ATRX) at pericentromeric hetrochromatin and the short arms of acrocentric chromosomes. *Proc. Natl. Acad. Sci. USA,* 1999, 96, 24, 13983-13988.

Medvedev Z. An attempt at a rational classification of theories of aging. *Biol. Rev.,* 1990, 65, 337-394.

Merington M., Penrose L. Distances which involve satellited chromosomes in metaphase preparations. *Ann. Hum. Genet.,* 1964, 27, 257-259.

Metz C. Chromosome studies on the Diptera II. The paired association of chromosomes in the Diptera, and its significance. *J. Exp. Zool.,* 1916, 21, 213-279.

Mier P., Kerkhof P. Programmed cell death in the epidermis. *Trends Biochem.,* 1990, 15, 3, 97-97.

Migliore L., Parrini M., Biagini C., et al. Hypothesen des Alterns-Stand und Probleme. Z. *Alternsforsch.,*1986, 2, 67-77.

Mikhelson V. Cell aging in cultures. Relation with natural aging and DNA reparation. Reliab. Element. Events of. Processes *Biol. Aging.,* Kiev, 1986, 116-123.

Miles C. Peripheral position of sex chromatin. *Natura,* 1961, 191, 7788, 626-627.

Miller D., Tantravahi R., Dev V., Miller O. Frequency of satellite association of human chromosomes is correlated with amount of Ag-staining of the nucleolus organizer region. *Amer. J. Hum. Genet.,* 1977, 29, 490-502.

Miller O., Breg W., Mukherjee B., et al. Non-random distribution of chromosomes in metaphase figures from cultured human leukocytes. II. The peripheral location of chromosomes 3, 17-18, 21. *Cytogenetics,* 1963b, 2, 2,-3, 152-167.

Miller O., Miller D., Dev V., Tantravahi R., Croce C. Expression of human and suppression of mouse nucleolus organizer activity in mouse human somatic cell hybrids. *Proc. Natl. Acad Sci. USA,* 1976, 73, 4531-4535.

Miller O., Mukherjee B., Breg W., van Gamble A. Non-random distribution of chromosomes in metaphase figures from cultured human leukocytes. The peripheral location of the Y-chromosomes. *Cytogenetics,* 1963a, 2, 1-14.

Milne C. An estimate of the heritability of worker longevity of length of life in the honeybee. *J. Apicult. Res.,* 1985, 24, 3, 140-143.

Mirsky A., Silverman B., Panda N. Blocking by histones of accessibility to DNA in chromatin. Addition of histones. *Proc. Nat. Acad. Sci. USA,* 1972, 69, 11, 3243-3246.

Monaselidze J., Chanchalashvili Z., Mgeladze G., et al. Thermal properties of intactnucleoprotiens. *J. Pol. Sci.,* 1981, 69, 17-20.

Monakhova M. Positional effect in the interphase nucleus. *Aging Cytogenetics. IV Intern. Symp., 26-29, 4, Tbilisi, 1989,* 28-29.

Monakhova M., Steppe Z. The centromeric association of metaphase chromosomes of white laboratory mice. *Cytologia i Genetika,* 1975, 9, 1, 45-47. (In Russian)

Moor K., Barr M. Nuclear morphology, according to sex in human tissues. *Acta Anat.,* 1954, 21, 6, 197-208.

Moorhead P., Nowell P., Mellman W., et al. Chromosome preparations of leukocytes cultured from human peripheral blood. *Exp. Cell. Res.,* 1960, 20, 3, 613-616.

Morgan W., Crossen P. The incidence of sister chromatid exchanges in cultured human lymphocytes. *Mut. Res,.* 1977, 42, 2, 305-312.

Morris I. Information and clonal concepts in aging. *Medical Hypothesis,* 1994, 42, 2, 89-94.

Mosgoller W., Leiten A., Brown F., et al. Chromosome arrangements in human fibroblasts at mitosis. *Hum. Benet.,* 1991, 88, 1, 27-33.

Mosgoller W, Lettch A., Brown J., Heslop-Harrison J.. Chromosome arrangements in human fibroblasts at mitosis. *Human Genetics,* 1991, 88, 27-33.

Moszhukhina T. *Age peculiarities of chromatine in tissues with different mitotic activities.* Author's abstract (of dissertation), Kiev, 1984, 1-24.

Muggleton-Harris A., Defaria R. Age-dependent metabolic changes in cultured human fibroblasts. *In vitro,* 1985, 21, 5, 271-276.

Mukherjee A., Weinstein M. Sequence of centromere separation of mitotic chromosomes during human cellular aging. *Mech. Aging and Dev.,* 1988, 45, 1, 59-64.

Muller W., Walkey D., Hager G., Mc Nally F. Large-scale chromatin decondensation and recondensation regulated by transcription from a natural promotor. *F. Cell Biol.,* 2001, 154, 1, 33-48.

Muramatsu S., Hanada H., Himenok. Effects of cadmium in biological tissues by atomic absorption induction in bonne marrow cells and spermatogenia on mice. *Radiobiological Equivalents. Chem. Pollit. Proc. Advis Group Meet.,* 1980.

Murthy D., Murthy S., Patel J., Tyagi A. Nucleolar organizer region heteromorphism associated with trisomy 21: a risk factor for non-disfunction. *Indian J. Exp. Biol.,* 1989, 27, 10, 864-867.

Nagl W. Chromatin organization and the control of gene activity. *Int. Rev Cytol,* Pt. A-Pt. B., 1985, 94, 21-56.

Nagele R., Freeman T, Mcmorrow L., Lee H. Precise spatial positioning of chromosomes during prometaphase: evidence for chromosome order. *Science,* 1995, 270: 1831-1835.

Nakagama T., Takami Y. Participation of histones and histone-modifying enzymes in cell functions through alterations in chromatin structure.J. Biochem.. 2001, 129, 491-499.

Nakagome Y., Abe T. The "loss" of kinetochores from chromosomes of aged individuals. *Ann. Rept. Inst. Genet.,* Japan, 1981 (1982), 32, 110-112.

Nakagome Y., Abe T., Misawa S., Takeshita T., Linuma K. The "loss" of centromeres from chromosomes of aged women. *Amer. J. Hum. Genet.,* 1984, 36, 2, 398-404.

Nakagome Y., Kitagawa T., Linuma K., et al. Pitfalls in the use of chromosome variants for paternity dispute cases. *Hum. Genet.,* 1977, 37, 3, 255-260.

Nakanishi Y., Kram D., Schneider E. Aging and sister chromatid exchange. IV. Reduced frequencies of mutagen induced sister chromatid exchanges in vivo in mouse bone marrow cells with aging. *Cytogenet. and Cell Genet.,* 1979, 24, 1, 61-67.

Natarajan A.,Klasterska I. Heterochromatin and chromatid exchanges in the chromosomes of Microtus agrestis. *Hereditas,* 1975, 79, 1, 150-154.

Nath J., Tucker J., Hande J. Y-chromosome aneuploidy, micronuclei, kinetochores and aging in men. *Chromosoma,* 1995, 103, 725-731.

Navashin M. Arrangement of metaphase chromosomes and nucleus dynamics. *Dokl. Acad. Nauk SSSR,* 1947, 7 ,6, 613-616. (In Russian)

Navashin S. About the arranging pair combination of chromosomes during the somatic cell division. *Dokl. Acad. Nauk SSSR,* 1926, 5, 142-144. (In Russian)

Nemeth A., Langst G. Chromatin higher order structure: Opening up chromatin for transcription. Brief. Funct. *Genomic Proteomic.* 2004, 2, 4, 334-343.

Nette E., Gerda Xi Yu-Ping, Sun Yu-Kay, et al. A correlation between aging and DNA repair in human epidermal cells. *Mech. Aging and Dev.,* 1984, 24, 3, 283-292.

Neurath P., DeRemer K., Bell B., Jarvik L., Kato T. Chromosome loss compared with chromosome size, age and sex of subjects. *Nature,* 1970, 225, 281-282.

New Haven Conference First intern. *Workshop on Human Gene Mapping. Birth Defects. Original Article Series,* 1973, X, 3, 55-57.

Nielsen J. Chromosomes in senile, presenile, and arteriosclerotic dementia. *J. Gerontol,* 1970, 25, 312-315.

Nisanov V., Volkov D., Saralidze E., et al. The system of automatic measurements of chromosome images of films. *Cytologia,* 1981, 23, 1, 109-112. (In Russian)

Nordenson L., Beckman G., Beckman L., Nordström S. Occupational and environmental risks in and around a smelter in northern Sweden IV. Chromosomal aberrations in workers exposed to lead. *Hereditas,* 1978, 88, 263-267.

Nuzzo F., Lagomarsini P., Casati A., et al. Clonal chromosome rearrangements in a fibroblasts strain from a patient affected by Xeroderma pigmentosum complementation group C. Mutat. Res. DNAging: *Genet. Instab. and Aging,* 1989, 219, 4, 209-215.

Ockey C. The position of chromosomes at metaphase in human fibroblasts and their DNA synthesis behaviour. *Chromosome,* 1969, 27, 308-320.

Ohmura H., Oshimura M. Telomere. cellular senescence and transformation. *Jap. J. Clin. Med.,* 1993, 51, 7, 1899-1906.

Ohno S., Trujillo J., Kaplan W., Kinozita P. Nucleolus-organisers in the causation of chromosomal anomalies in man. *Lancet,* 1961, 2, 123-126.

Ohtaki K., Sposto R., Kodama Y., et al. Aneuploidy in somatic cells of in utero exposed A-bomb survivors in Hiroshima. *Mut. Res.,* 1994, 316, 1, 49-58.

Olins A., Olins D. Spheroid chromatin units (v-bodies). *Science,* 1974, 83, 330-332.

Olson M., Dundr M., Szebeni A. The nucleolus: an old factory with unexpected capabilities. *Trends Cell Biol.,*2000, 10, 5, 189-196.

Orey E. Satellite association and variations in the length of the nucleolar construction of normal and variant human G chromosome. *Humangenetik,* 1974, 22, 299-309.

Orgel L. The maintenance of the accuracy of protein synthesis and its relevance to aging. *Proc. Natl. Acad. Sci. USA,* 1963, 49, 517-521.

Orgel L. The maintenance of the accuracy of protein synthesis and its relevance to aging accorrection. *Proc Natl. Acad. Sci USA,* 1970, 67, 1476-1476.

O'Riordan M,. Berry E. Tough I. Chromosome studies on bone marrow from a male control population. *Brit. J. Hematol.,* 1970, 19, 83-90.

Painter R., Clarkson J., Young B. UV-induced repair replication in aging diploid human cells (WI-38). *Rad. Res.,* 1973, 56, 560-564.

Painter T., Reindorp E. Endomitosis in the nurse cells of the ovary of Drosophila melanogaster. *Chromosoma,* 1939, 1, 2, 276-283.

Pandata T., Pathak S., Geard C. Chromosome end associations, telomeres and telomerase activity in ataxia telangiectasia cells. *Cytogenet. Cell Genet.,* 1995, 71, 1, 86-93.

Panno J., Nair K. Chromatin condensation in the aging housefly. *Exp. Gerontol.,* 1984, 19, 1, 63-72.

Pardue M., Gall J. Chromosomal localization of mouse satellite DNA, *Science,* 1970, 168, 1356-1358.

Paris Conference. Standardization in human cytogenetics. *Birth Defects Orig. Artic, Ser., VIII.* New York, 1972.

Park W. The occurrence of sex chromatin in early human and macaque embryos. *J. Anat.,* 1957, 91, 3, 369-373.

Patil S., Lubs H. Non-random association of human acrocentric chromosomes. *Humangenetik,* 1971, 13, 157 159.

Patil S., Lubs H. Classification of 9h regions in human chromosomes 1, 9, 16 by C-banding. *Hum. Genet.,* 1977, 38, 35 –38.

Pennisi E. Premature aging gene discovered. *Science,* 1996, 272, 5259, 193-194.

Penrose L. Paternal age in Mongolism. *Lancet,* 1962, 1, 7239, 1101-1101.

Perry P., Wolff Sh. New Giemsa method for the differential staining of sister chromatids. *Nature,* 1974, 251, 5481, 156-158.

Perry R. Nucleoli: The cellular sites of ribosome production. In: Lima de Faria (Ed.). *Handbook of Molecular Cytology.* Amsterdam-London. North Holland Publishing Company, 1969, 620-636.

Phillips R. Inheritance of acrocentric association patterns. *Humangenetik,* 1975, 29, 309-318.

Pierre R., Hoagland H. Age-associated aneuploidy: loss of Y-chromosome from human bone marrow cells with aging. *Cancer,* 1972, 30, 4, 889-894.

Pile L., Wassarman D. Chromosomal localization links the SIN3-RBD3 complex to the regulation of chromatin condensation histone acetylation and gene expression. *EMBO,* 2002, 19, 22, 6131-6140.

Pilins'ka M., Dybs'kyi S. FISH detection of spontaneous level of chromosome aberration in peripheral blood lymphocytes in persons of different ages. *Tsitol. Genet.* 2004, 38, 4, 62-66.

Pincheira J., Gallo C., Brano M., et al. Repair and aging: influence of donor age on chromosomal aberrations in human lymphocytes. *Mut. Res.,* 1993, 295, 2, 55-62.

Platz E., Helzisouer K., Hoffman S., et al. Prediagnostic toenail cadmium and zinc and subsequent prostate cancer risk. *Prostate,* 2002, 52, 4, 288-296.

Podugolnikova O., Korostelev A. The quantitative analysis of polymorphism on human chromosome 1, 9, 16 and Y. IV. Heterogeneity of a normal population. *Hum. Genet.,* 1980, 54, 2, 163-169.

Potapenko A., Akifiev A. On the way of the program and initial substrate of aging. *Adv.Gerontol.,*1999, 3, 68-80.

Prashad N., Cutler R. Percent satellite DNA as a function of tissue and age of mice. *Biochem. and Biophys. Acta,* 1976, 418, 1-23.

Prescott D., Myerson O., Wallace J. Enucleation of mammalian cells with cytochalasin B. *Exp. Cell Res.,* 1972, 71, 480-485.

Prieur M., Al Achkar W., Aurias A., et al. Acquired chromosome rearrangement in human lymphocytes effect of aging. *Hum. Genet.,* 1988, 79, 2, 147-150.

Probst A., Fransz P., Paszkowski J., Sched O. Two means of transcriptional reactivation within heterochromatin. *Plant J.* 2003, 33, 4, 743-749.

Prokofyeva-Belgovskaya A. *Heterochromatin regions of chromosomes.* M., Nauka, 1986, 3-431.

Prokofieva-Belgovskaya A., Gindilis V. Identification of the human chromosomes. *Izvestia Akademii Nayk ,* B,1965,2,188-200

Raes M., De Brabander M., Remacle J. Polyploid cells in aging hamster fibroblasts in vitro: possible implication of the centrosome. *Eur. J. Cell Biol.,* 1984, 35, 1, 70-80.

Ramsey M., Moore D., Briner J., et al. The effects of age and lifestyle factors on the accumulation of cytogenetic damage as measured by chromosome paiting. *Mutat. Res.,* 1995, 338, 1 6.

Reff M., Schenieder E. Cell culture aging. *Mol. and Cell Biochem.,* 1981, 36, 3, 169-176.

Regan J., Letlor R., Lay R. Normal and defective repair of damaged DNA in human cells. A sensitive assay utilizing the photolysis of bromodeoxyuridine. *Proc. Natl. Acad. Sci., USA,* 1971, 68, 708-712.

Remacle J. Theories of cellular aging. *Arch. Biol.,* 1985, 96, 2, 237-237.

Revell S. A new Hypothesis for chromatid changes. In: *Proc. Radiobiol. Symp. Liege,* 1954, Betterworth, London, 1955, 243-253.

Riabewol K., Shmokler R., Goldstein S. Interspersed repetitive and tandemly repetitive sequences are differentially represented in extrachromosomal covalently closed circular DNA of human diploid fibroblasts. *Nucl. Acids Res.,* 1985, 13, 15, 5563-5584.

Richard F., Muleris M,. Dutrilaux B. The frequency of micronuclei with X-chromosome increases with age in human females. *Mut. Res.,* 1994, 316, 1, 1-7.

Ris H. The structure of chromosome. *A symposium on the chemical basis of heredity.* Baltimore, 1957.

Roa P., Johnson R. Induction of chromosome condensation in interphase cells. In: *Advances in cell and molecular biology* (ed. E.J. Du Praw). New York Acad. Press, 1974, 3, 135-189.

Robbins W., Baulch J., Moore O., et al. Three-probe fluorescence in situ hybridization to assess chromosome X, Y and 8 aneuploidy in sperm of 14 man two healthy groups: evidence for a paternal age effect on sperm aneuploidy. *Reprod. Fertil Dev.,* 1995, 7, 4, 799-809.

Rockstein M., Chesky J., Gassman M. *Comparative biology and evolution of aging.* (Eds.). C. Finch and L. Hayflick, Van No-strand-Reinholds. New York, 1977, 1-34.

Ronne M. Human cells in Suspension 2. The effect of in vitro aging on metaphase chromosome structure in human lymphoid cells. *Hum. Genet.,* 1980, 54, 55-62.

Roth M., Emmons L., Haner M. Age related decrease in an early step of DNA-repair of normal human lymphocytes exposed to ultraviolet irradiation. *Exp. Cell, Res.,* 1989, 1, 171-177.

Ryan J., Cristofalo V. Chromatin template activity during aging in WI-38 cells. *Exp. Cell Res.,* 1975, 90, 456 456.

Saksela E., Moorhead P. Aneuploidy in the generative phase of serial cultivation of human cell strains. *Proc. Nat. Acad. Sci. USA,* 1963, 50, 390-394.

Saadat I., Allamenh A., Saadat M. DNA-repair capacity in Down"s syndrome. *Iranian Biomed. J.,*1998, 2, 123-127.

Sandberg A., Cohen M., Rimm A., Levin M. Aneuploidy and age in a population survey. *Amer. J. Hum. Genet.,* 1967, 19, 5, 633-643.

Santoro R., Li J., Grummt I. The nucleolar remodeling complex NORC mediates heterochromatin formation and silencing of ribosomal gene transcription. *Nat. Genet.,* 2002, 32, 3, 393-396.

Saralidze E., Ulyanov N., Pogorelov V., Kaminir L. The measurement of C-banded chromosomes. *Cytologia,* 1981, 23, 5, 599-560. (In Russian)

Schmiady H., Munke M., Sperling K. Ag-staining of nucleolus organizer regions on human prematurely condensed chromosomes from cells with different ribosomal RNA gene activity. *Exp. Cell Res.,* 1979, 121, 2, 425-428.

Schmid M., Haaf T., Schindler D., Meurer M. Centromeric association of a microchromosome. A new category of non-random arrangement of metaphase chromosomes. *Hum. Genet.,* 1989, 81, 2, 127-136.

Schmid M., Krone W., Vogel W. On the relationship between the frequency of association and the nucleolar constriction of individual acrocentric chromosomes. *Humangenetik,* 1974, 23, 267-277.

Schneider E., Kram D., Nakanishi Y., Monticone R. The effect of aging on sister chromatid exchanges. *Mech. Aging and Dev.,* 1979, 9, 3-4, 303-311.

Schneider E., Mitsui Y., Au K., Shore S. Tissue specific differences in cultured human diploid fibroblasts. *Exp. Cell Res.,* 1977, 108, 1-6.

Schneiderman L., Smith C. Non-random distribution of certain homologous pairs of normal human chromosomes in metaphase. *Nature,* 1962, 195, 4847, 1229-1230.

Schroder H., Bernd A., Zahn R., Muller W. Age-dependent alterations of microtubule-associated enzyme activities from bovine brain. (Protein kinase, adenosine triphosphatase, guanosine triphospatase). *Mech. Aging and Dev.,* 1983, 22, 35-50.

Schwartzman J., Cortes F. Sister chromatid exchanges in Allium cepa. *Chromosoma,* 1977, 62, 119-131.

Schwartz D. Evidence for sisters strand crossingover in maire. *Genetics,* 1953, 38, 251-260.

Schwartzacher H., Schnedl W. Endoreduplication in human fibroblast cultures. *Cytogenetics,* 1965, 4, 1, 2-18.

Seabright M,. Gregson N., Mould S. Trisomy 9 associated with an enlarged 9 qh segment in aliveborn. *Hum. Genet.,* 1976, 34, 3, 323-325.

Sele B., Jalbert P., van Cutsem B., et al. Distribution of human chromosomes on the metaphase plate using banding techniques. *Hum. Genet.,* 1977, 38, 39-61.

Setlow R.,Carrier W. The disappearance of thymidine dimers from DNA : An error correcting mechanism. *Proc. Nat. Acad. Sci. USA,* 1964, 51, 226-231.

Sevankaev A., Kozlov V., Gusev G,. Ismailova N. Frequency of spontaneous chromosome aberrations in the culture of human leukocytes. *Genetika,* 1974, 10, 6, 114-120. (In Russian)

Sharma A., Totukdar G. Effects of metals on chromosomes of higher organisms. *Environ. Mutag.,* 1987, 9, 191-226.

Shekhter-Levin S., Sherer M., Wald N., Gollin S. Loss of the Y-chromosome from bone marrow cells. Is it an ageing phenomenon or a malignancy marker? Cytogenet. *Cell Genet.,* 1992, 60, 3-4, 269-269.

Shen G., Galick H., Inoue M.,Wallace S. Decline of nuclear and mitochondrial oxidative base excision repair activity in late passage human diploid fibroblasts. *DNA repair,* 2003, 2, 6, 673-693.

Shmookler Reis R., Lumpkin J., McGill R., et al. Extrachromosomal circular copies of an “inter-Alu” unstable sequence in human DNA are amplified during in vitro and in vivo aging. *Nature,* 1983, 301, 394-398.

Sigmund J., Schwartzacher H., Mikelsaar A. Satellite association frequency and number of nucleoli depend on cell cycle duration and NOR-activity. *Hum. Genet.,* 1979, 50, 1, 81-91.

Simi S., Tursi F. Polymorphism of human chromosomes 1, 9, 16, Y variations, segregation and mosaicism. *Human Genet.,* 1982, 62, 3, 217-220.

Singh R., Robbelen G., Okanoto M. Somatic association at interphase studied by Giemsa banding technique. *Chromosoma,* 1976, 56, 265-273.

Small M., Aronson M. Age related changes in excision repair of cultured epithelial cells from mouse thymus. *Mech. Aging and Dev.,* 1988, 45, 2, 127-136.

Smith D., Evans H. Mapping of SCE in human chromosome, using G-banding and autoradiography. *Mut. Res.,* 1976, 35, 1, 139-154.

Smith J. Human Sir2 and the "silencing of p53 activity". *Trends Cell Biol.,* 2002, 12, 9, 404-406.

Smith F., Caputo F., Bocke F. A genetic screen for rDNA silencing defects medicates multiple PNH replication and chromatin modulating factors. *Mol. Cell Biol.* 1999, 19, 4, 3184-3197.

Smith M., Bortolotto M., Melaragno M., Toniolo N. Investigation of the effect of hydrogen peroxide on the chromosome of young and elderly individuals. *Mech. Aging and Dev.,* 1990, 56, 2, 107-115.

Soewarto P., Schmiady H., Eichenlaubritter U. Consequences of non-extrusion of the first polar body and chromatids in mammalian oocytes. *Hum. Reproduction,* 1995, 10, 9, 2350-2360.

Sokolov I. Chromosomes in the spermatogenesis of domestic donkey. *Dokl. Acad. Nauk SSSR,* 1937, 15, 5-7, 361-364. (In Russian)

Solomon E., Bobrow M. Sister chromatid exchanges - a sensitive assay of agents damaging human chromosomes. *Mut. Res.,* 1975, 30, 2, 273-278.

Sosa M., Cories F. Chromosome arrangement throughout mitosis and interphase in Allium sativum. *Experientia,* 1984, 340, 3, 285-286.

Sperling K., Wegner R., Riechm H., Obe G. Frequency and distribution of sister chromatid exchanges in a case of Fanconi's anemia. *Humangenetik,* 1975 27, 3, 227-230.

Spiegler P., Norman A. Temperature dependence of unscheduled DNA synthesis in human lymphocytes. *Radiat. Res.*, 1970, 43, 187-195.

Stahl A., Luciani J. Nucleoli and chromosomes. Their relationships during the meiotic prophase of the human fetal oocyte. *Humangenetik,* 1972, 14, 269-284.

Stahl A., Luciani J., Nicoling J., et al. Maladie de klinefelter a mosaique XY-XXY: Double satellite sur le chromosome 13. *Ann. Endocr.,* 1966, 27, 147-154.

Staino-Coico L., Darzynkiewicz Z., Hefton J., et al. Increased sensitivity of lymphocytes from people over 65 to cell cycle arrest and chromosomal damage. *Science,* 1983, 219, 4590, 1335-1337.

Stefanov S. Random-step morphometric grid for rapid measurement of morphogenetic elements. *Cytologia,* 1974, 16, 6, 785-787. (In Russian)

Stern C. Somatic crossingover and segregation in Drosophila melanogaster. *Genetics,* 1936, 21, 625-730.

Stevens N. A study of the germ cells of certain Diptera, with reference to the heterochromosomes and the phenomenon of synapses. *J. Exp. Zool.,* 1908, 5, 359-374.

Stewart J. Genetic factors on the X-chromosome. *Lancet,* 1961, 2.7197, 317-317.

Stitou S., Diaz de la Guardia R., Jimenes R., Burgous M. Inactive ribosomal cistrons are spread throughout the B chromosomes of Rattus (Rodentia Muridal) implactions for their origin and evolution. *Chromosome Res.,* 2000, 8, 305-311.

Stukalov S. Differential Repair Activity of human chromosomes. *Bull. Exper. Biol. and Med.,* 1995, 119, 1, 63 65. (In Russian)

Sumner A. A simple technique for demonstrating centromeric heterochromatin. *Exp. Cel. Res.,* 1972, 75, 304 306.

Sumner A. Banding as a level of chromosome organization. In: *Current Chromosome Research* (K. Jones, P. Brandham eds.) Amsterdam, Elsevier - North Holland, Biochemical Press, 1976, 17-22.

Sumner A., Evans N., Buckland R. New technique for distinguishing between human chromosomes. *Nature New Biol.,* 1971, 232, 31-31.

Sun H., Shen J., Yokota H. Size-dependent positioning of human chromosomes in interphase nuclei. *Biophis. J.,* 2000, 79, 184-190.

Suzuki T, Fujii M, Ayusawa D. Demethylation of classical satellite 2 and 3 DNA with chromosomal instability in senescent human fibroblasts.Exp *Gerontol.,* 2002, 37,8-9,1005-1014.

Sutnick A., London W., Blumberg B., Gerstley B. Susceptibility to leukemia immunologic factors in Down's syndrome. *J. Nat. Cancer Inst.,* 1971, 47, 923-928.

Suzuki T., Fuji M., Ayu Sawa D. Demethylation of classical satellite 2 and 3 DNA with chromosomal instability in senescent human fibroblasts. *Exp. Gerontol.,* 2002, 37, 8-9, 1005-1014.

Svidchenko A. Somatic association of chromosomes. *Cytologia i Genetika,* 1975, 9, 5, 464-470. (In Russian)

Szweda P., Camouse M., Lundberg K., Oberley T., Szweda L. Aging, lipofuscin formation, and free radical-mediated inhibition of cellular proteolytic systems. *Ageing Res. Rev.* 2003, 2, 4, 383-405.

Tada-Aki Hori. Induction of chromosome decondensation, sister chromatid exchanges and endoreduplications by 5-azocytidine an inhibitor of DNA methylation. *Mut. Res.,* 1983, 121, 47-52.

Takeshita T., Ariizumi-Shibusawa C., Shimizu K., et al. The effect of aging on cell-cycle kinetics and X-ray-induced chromosome aberrations in cultured lymphocytes from patients with Down syndrome. *Mutat. Res.,* 1992, 1, 21-29.

Tamayo J. Structure of human chromosomes studied by atomic force microscopy. *J.Struct. Biol.,* 2003a, 141, 3, 198-207.

Tamayo J. Structure of human chromosomes studied by atomic force microscopy. Part II. .Relationship between structure and cytogenetic bands. *J.Struct. Biol.,* 2003b, 141, 3, 189-197.

Tang D.,Hwang B.,Ford J.,et al. Xeroderma pigmentosum p48 gene enhances global genomic repair and suppresses UV-induced mutagenesis. *Mol. Cell,* 2000, 5, 4, 737-744.

Tantravahi R., Miller D., Miller O. Ag-staining of nucleolus organizer regions of chromosomes after Q-, C-, G- or R-banding procedures. *Cytogenet. and Cell Genet.,* 1977, 18, 364-469.

Tarin J. Etiology of age-associated aneuploidy. A mechanism based on the free-radical theory of aging. *Hum. Reproduction,* 1995, 10, 6, 1563-1565.

Taylor E., Martin-De Leon P. Familial silver staining patterns of human nucleolus organizer regions (NORs). *Amer. J. Hum. Genet.,* 1981, 33, 1, 67-76.

Taylor J. Sister chromatid exchanges in tritium labeled chromosomes. *Genetics,* 1958, 43, 515-529.

Terman A., Brunk U. Lipofuscin. *Int. Biochem. Cell Biol.,* 2004, 36, 8, 1400-1404.

Thierens H., Vral A., De Ridder L. A cytogenetic study of radiological workers: effect of age, smoking and radiation burden on the micronucleus frequency. *Mutat. Res.,* 1996, 360, 2, 75-82.

Theophanides T., Anastassopoulou J. Cooper and carcinogenesis. Crit. Rev. *Oncol. Hematol.,* 2002, 42, 1, 57-64.

Thoma F. Light and dark in chromatin repair: of UV - induced DNA lesions by photolyase and nucleotide excision repair. *EMBO Journal.* 1999, 18, 6585-6598.

Tice R., Chaillet G., Schneider E. Evidence derived from sister chromatid exchanges and restricted rejoining of chromatid submits. *Nature,* 1975, 256, 642-644.

Tjio J., Puck T., Robinson A. The human chromosomal satellites in normal persons and in two patients with Marfan's syndrome. *Proc. Natl. Acad. Sci, USA,* 1960, 46, 532-539.

Tone S.,Tanaka S., Kato Y. The cell cycle and cell population kinetics in the progrommed cell death in the limb-buds of normal and 5-bromdeoxyuridine-treated chik embryos. *Dev.Srouth. and Differ.,* 1988,30,3,261-270.

Tongh I., Court Brown W., Baikie A., et al. Cytogenetic studies in chronic myeloid leukemia and acute leukemia with Mongolism. *Lancet.,* 1961, 1, 411-411.

Touchette N.,Cole D. Differential Scanning Calorimetry of nuclei reveals the loss of structural features in chromatin by brief nuclease treatment. *Natl. Acad. Sci. USA,* 1985,82, 2642-2646.

Toussaint O., Medrano E., von Zglinicki T. Stress-induced premature senescence (SIPS) of human diploid fibroblasts and melanocytes. *Exp. Gerontol.,* 2000, 35, 927-945.

Toussaint O., Royer V., Salmon M., Ramacle F. Stress-induced premature senescence and tissue ageing *Biochemical Pfarmacology,* 2002, 64, 1007-1009.

Trere D. Ag-NOR staining and quantification. *Micron,* 2000, 31, 127-131.

Tsoneva M., Krashunova M., Tsansheva M., Lazanova T. C-band polymorphisms of chromosome Y- association with ageing. *Sovremen. Med.,* 1980, 31, 8, 432-435.

Tumilovich L., Lezhava T. Clinico – cytogenetic studies in dysgenesis gonads. *Akusherstvo i Ginekologia,* 1968, 4, 19-29. (In Russian)

Turjin R., Lejeune J. *Les chromosomes humains gauthiervillars,* Paris, 1965.

Turner B.,Koehane A. Cu2+ - dependent changes in metaphase chromosomes structure asseyed by antibody labelling and flow cytometry. *Biochem. Soc. Trans.,* 1986, 4, 6, 1166-1167.

Uchida I., Lee C., Byrnes E. Chromosome aberrations induced in vitro by low of radiation: Nondisjunction in lymphocytes of young adults. *Amer. J. Hum. Genet.,* 1975, 27, 3, 419-429.

Ura K., Araki M., Sneki H., Masutuni C., Ito T., Iwai. S., Mizukoshi T., Kaneda Y., Hanaoka F. ATP-dependent chromatin remodeling facilitates nucleotide excisian repair of UV-induced DNA-lesions in synthetic dinucleosomes. *EMBO J.* 2001, 20, 2004-2014.

Ueda N., Uenaka H., Akematsu T,. Sugryama T. Parallel distribution of SCE and chromosome aberrations. *Nature,* 1976, 262, 5569, 581-583.

Uhlmann F. Chromosome condensation: Packaging the genome. *Current Biology,* 2001, 11, 10, R384 - R387.

Van Dyke D., Palmer C., Nance W., Yu P. Chromosome polymorphism and twin zygote. *Amer. J. Hum. Genet.,* 1977, 29, 431-447.

Van-Ganzen P. Cell aging in vitro. *Cytologia,* 1979, 21, 6, 627-649. (In Russian)

Vaniushin B., Romanenko E. DNA methylation and genome structure at animal aging. In: *Artificial increase of specific lifetime.* 8-10, 12, Moskva, 1980, 9-10.

Vaquero A., Scher M., Lee D., Erdjument – Bromage H., Tempst P., Reinberg D. Human SirT1 interacts with histone H1 and promotes formation of facultive heterochromatin. *Mol. Cell,* 2004, 16, 93-105.

Vaquero A., Loyola A., Reinberg D. The constantly changing face of chromatin- Sci. *Aging Knowledge Environ,* 2003, 14, ref 4, 1-16.

Vaziri H., Schachter F., Uchida J., et al. Loss of telomeric DNA during aging of normal and trisomy 21 human lymphocytes. *Amer. J. Hum. Genet.,* 1993, 52, 4, 661-667.

Verma R., Lubs H. Variations in human acrocentric chromosomes with acridine orange reverse banding. *Humangenetik,* 1975, 30, 225-235.

Verma R., Rodriguez J. Structural organization of ribosomal cistrons in human nucleolar organizing chromosomes. *Cytobios,* 1985, 44, 175, 25-28.

Verma R., Shah J., Dosic H. Frequencies of chromosome and chromatid types of associations of nucleolar human chromosomes demonstrated by the N-banding technique. *Cytobios,* 1983, 36, 25-29.

Verschaeve L., Kirsch-Volders M., hens L., Susanne C. Chromosome distribution and mercury. *Mut. Res.,* 1978, 53, 2, 279-279.

Vig B. Sequence of centromere separation occurrence, possible significance, and control. *Cancer Genet. and Cytogenet.,* 1983, 8, 249-274.

Vijg J., Dolle M. Larg genome rearrangements as a primary cause of aging. *Mech. Ageing and Dev.,* 2002,123,8,907-915.

Vijg J., Knook D. DNA repair in relation to the aging process. *J. Amer. Geriatr. Soc.,* 1987, 35, 6, 532-541.

Vilenchik M. New Approaches to investigation of the role of DNA damage and reparation in aging in connection with the problem of spontaneous carcinogenic. In: *Molecular and cellular mechanisms of aging.* Kiev, 1981, 35-37.

Vormittag W. Effect of donor age on sister chromatid exchange frequency in cultured human lymphocytes. *Actual Gerontol.,* 1983, 13, 2, 79-81.

Vormittag W., Kubbock J., Meininger M. Vergroberungregion des Kurzen armes eines kleinen akrozentrischen chromosoms. *Hemangenetik,* 1972, 14, 2, 112-121.

Vorobtsova I., Kanaeva A., Petrova J., Semenov A., Pleskach N., Spivak M., Timonina G., Prokof'eva V., Partseva N., Mikhelson V. Age dynamics of stable chromosome aberration frequency in human with natural and pathological senescence. *Tsitologia,* 2004, 46, 12, 1030-1034.

Vorobtsova I., Semenov A., Timofeyeva a., Zvereva I. An investigation of the age-dependency of chromosome abnormalities in human populations exposed to low-dose ionizing radiation. *Mech. Ageing Dev.2001,* 122, 13, 1373-1382.

Walter S., Weiler J., Moschuring K. Chromosomal loss in both sexes of aged and castrated hamsters (Mesocricetus auratus) Abh., Akad. Wiss und Lit. Math-naturwiss Kl. *Res. Mol. Biol.,* 1981, 10, 118-124.

Warburton D. Biological aging and the of aneuploidy. *Cytogenetic Genome Res.* 2005, 111, 3-4, 266-272.

Warburton D., Naylor A., Warburton F. Spatial relations of human chromosomes identified byquinacrine fluorescence at metaphase. *Humangenetik,* 1973, 18, 297-306.

Warner H., Campisi L., Cristofalo V., et al. Control of cell proliferation in senescent cells. *J. Grontol.,* 1992, 47, 185-189.

Warner H., Sierra F. Models of accelerated ageing can be informative about the molecular mechanisms of aging and / or age-related pathology. *Mech. Ageing and Dev.,* 2003, 124, 5, 581-587.

Warwick M., Lawric S., Beveridge A., John S. Abnormal cerebral asymmetry and schizophrenia in a subject with Kleinfelter's syndrome(XXY).*Biol. Psychiatr.,* 2003, 53, 7, 627-629.

Watanabe T.,Shimada T.,Endo A. Mutagenic effects of cadmium on mammalian oocyte chromosomes. *Mutat. Res.,* 1979, 67, 344-349.

Weimer R., Haaf T., Kruger F., Poot M., Schmid M. Characterization of centromere arrangement and test for random distribution in G_0, G_1, S, G_2, G' and early S' phase in human lymphocytes. *Hum. Genet.,* 1992, 88, 6, 673-682.

Weirich-Schwaiger H., Weirich H., Gruber B., et al. Correlation between senescence and DNA repair in cells from young and old individuals and in premature aging syndromes. *Mutat. Res.,* 1994, 316, 1, 37-48.

Wen Wu-Nan, Liew Tai-Lin, Wu Shong WangJan. The effect of age and cell proliferation on the frequency of sister chromatid exchange in human lymphocytes cultured in vitro. *Mech. Aging and Dev.,* 1983, 21, 2-3, 377-384.

Werry P., Stroffelsen K., Engels F., et al. The relative arrangement of chromosomes in mitotic interphase and metaphase in Huplapappus gracilis. *Chromosoma,* 1977, 62, 1, 93-101.

Weirich-Schwaiger H., Weirich H., Gruber B., et al. Correlation between senescence and DNA repair in cells from young and old individuals and in premature aging syndromes. *Mut. Res.,* 1994, 316, 1, 37-48.

Wheeler K., Lett J. On the possibility that DNA repair is related to age in nondividing cells. *Proc. Nat. Acad. Sci. USA,* 1974, 71, 1862-1865.

Wickens A. P. Ageing and the free radical theory *Respir Physiol.,* 2001, 128, 3, 379-391.

Widlak P, Palyvoda O, Kumala S, Garrard W. Modeling apoptotic chromatin condensation in normal cell nuclei. Requirement for intranuclear mobility and actin involvement. *J. Biol Chem.,* 2002 , 277, 24, 21683- 21690.

Wilson A., Carison S., White T. Biochemical evolution. *Rev. Biochem.,* 1977, 46, 573-639.

Woodruff R.,Nikitin A. P DNA element movement in somatic cells reduces in Drosophila melanogaster: evidence in support of the somatic mutation theory of aging. *Mutat. Res.,* 1995,338,1-6,35-42.

Wolff Sh. Are sister chromatid exchanges sister strand crossovers or radiation-induced exchanges *Mutat. Res.,* 1964, 1, 4, 337-343.

Wolff Sh. Sister chromatid exchanges. *Ann. Rev. Genet.,* 1977, 11, 183-201.

Wolff Sh., Bodycote J., Thomas G., Cleaver G. Sister chromatid exchanges in Xeroderma pigmentosum cells that are defective in DNA excision repair or post-replication repair. *Genetics,* 1975, 84, 2, 349-355.

Wolffe A., Hayes F. Chromatin dis ruptin and modifi eation. *Nucleic Acids Res.,* 1999, 27, 711-720.

Wong K., Hayes K. Leung S., Glassman A. The relationship between age and karyotypic abnormalities in myelodysplastic syndromes. *Cancer Genet. Cytogenet.,* 1996, 88, 1, 80-82.

Wright W., Hayflick L. Nuclear control of cellular aging demonstrated by hybridization of anucleate and whole cultured normal human fibroblasts. *Exp. Cell Res.,* 1975 96, 113-121.

Wu C. Chromatin remodeling and the control of gene expression. *J. Biol. Clin.* 1997, 272, 45, 28171-28174.

Xiao Y.,Tates J., Boel A., Natarajan A. Aging and diethylstibestrol-induced aneuploidy in male germ cells: a trangenic mouse mutational model. *Chromosoms,* 1998, 107, 507-513.

Yang J., Anzo M., Cohen P. Control of aging and longevity by AGF-I signaling. *Exper. gerontol.* 2005, 40, 11, 867-872.

Yielding K. A model for aging based on differential of somatic mutational damage. Perspect. *Biol. Med.,* 1974, 17, 2, 201-208.

Yu C., Borgaonkas O. Normal rate of SCE in Down syndrome. *Clinical Genet.,* 1977, 11, 6, 397-401.

Zakharov A. Sister chromatid exchanges: phenomen and mechanisms. In: *Human Genetics.* L., Meditsina, 1978, 114-161.

Zaks L Statistical appraise. M., *Statistic,* 1976, 576.

Zang K., Back E. Quantitative studies on the arrangement of the human metaphase chromosomes 1. Individual features in the association pattern of the acrocentric chromosomes of normal males and females. *Cytogenetics,* 1968, 7, 455-470.

Zang K., Zankl H. Chromosome size and hypodiploidy. *Nature,* 1970, 228, 778-779.

Zankl H., Zang K. Quantitative studies on the arrangement of human metaphase chromosomes IV. The association frequency of human acrocentric marker chromosomes. *Humangenetik,* 1974, 23, 259-265.

Zanzoni K., Baumann J., Jung E. Sister chromatid exchange (SCE) in human lymphocytes. *Arch. Derm. Cytol. Res.,* 1979, 265, 3, 283-287.

Zasukhina G., Sinelchikova T. Repair processes in mammalian cells. *Adv. Modern Genet.,* 1979, 8, 84 96. (In Russian)

Zdzienicka M., Verhaegh G., Jongmans W., et al. Functional Complementation studies with X-ray-sensitive mutants of Chinese Hamster cells closely resembling Ataxia-Telangiectasia cells. *Int. J. Radiat. Biol.,* 1994, 66, 6, S189-S195.

Zellweger H., Abbo G. About a new gene as a cause of increased satellite association. *J. Ped.,* 1965, 67, 2, 5-5.

Zhestianikov V. *DNA reparation and its biological importance.* L., Nauka, 1979.

Zhestianikov V., Vilenchik M. DNA reparation and aging. In: *Problems of radiation gerontology.* M., Atomizdat, 1978, 23-28.

Zhivago P. About the chromosome complex of turkey-hen. *Exper. Biol.* (ser.A), 1928, 4, 2, 215-225. (In Russian)

INDEX

A

B

C

D

E

F

I

J

K

L

M

N

O

P

Q

R

S

T

U

V

W

X

Y

Z